Mentally Discharged

Joe Rainer McClain Jr.

ISBN: 978-0-9996068-2-7

DEDICATION

To the five sailors and more importantly MEN who changed my life, and taught me more than I could ever imagine. Myles, Marcus, Frank Jackson, KJ and Fields. Thank you. Y'all did more for me than you all will ever know. Whatever y'all do in life, y'all will thrive. CG-52 fam.

CONTENTS

ACKNOWLEDGMENTS

To the Sailor, Marine, Airman, Soldier, Coast Guardsmen who are fighting a fight that no one can see. I am here to tell you that it is ok to not be ok. If anyone knows what you are going through, it is me. I acknowledge you here, hoping that the world will do the same.

FOREWORD

Dear Joe (From God),

Hey you. 17 year old knucklehead. It's June 2002 and it's hot in this recruiting office. You don't know it, but your parents unwillingness to get you down to FAMU (Florida A&M University, Historically Black College and University) is gonna be one of the biggest blessings ever. I saw your disappointment when Mrs. Hayes got you the scholarship after you dropped a 1020 on the SAT, and you told her you wouldn't be able to go. Hold on. That favorite teacher of yours, Mrs. Joshua. You'll understand why she had you write and do poetry consistently later. Oh, about this military stuff. Yea, I know you were hell bent on the Air Force, cause you were so upset at your father. But, I couldn't let you leave without some money. So, I gave you 10Gs to sign your name with the Navy. Now, you're gonna blow it, cause you're young, and don't listen. Don't worry. It's a part of the process. You're gonna leave for your second deployment with less than $100 to your name. You're gonna blow out your knee. You're gonna have a heart attack. You're gonna get married, then divorced. Hold on man. It gets worse. You're gonna lose your dad, your Godmother Jackie, your Uncle Chicken, your best Friend Tre and your sister Michelle in a three year span. The latter two will both die within 11 days of each other. Your relationship with your birth mother, well, you'll see. Lemme also add I'm gonna move you overseas, and have you get hit head on by a lifted F150. I could take you, but I'm gonna let you keep going. I'm gonna put you at a place for 4 years that is as bad as hell. Remember the name Harpers Ferry. Oh, I can't tell you everywhere you will go. Just know, YOU WILL get in trouble in Sydney, Australia. But, I'm gonna have your cousin save your career. Crazy y'all ain't gone know y'all cousins til y'all out there. You'll also have people turn their backs on you. Now, for some good news. It's gonna be some people along the way who gone ride with you. I'm talking ride! I can't give you their names, but they'll be from Texas, Oklahoma, Nashville, Arkansas and East St. Louis, Illinois. You'll meet some other hittas who'll prepare you for battle, but these five will go into war with you. I know this a lot for your 17 year old brain. Trust me though fam. On October 31st, 2022, I want you to read this, knowing I had a plan for you 20 years ago. Last but not least. I made burritos for everyone. Stop eating 'em all. You're looking like "Huh. I don't eat Mexican food like that." Trust me. Where I'm taking you, you'll be half Latino, you'll start saying relly instead of cuz and you will learn terrible driving skills, cutting off people on the freeway without a care in the world. Just wait on it.

Sincerely, God.

After 20 years of continual service, I've finally introduced myself, to myself again. He has changed, but that's what grown men do. If you still see me as the guy you grew up with, or the 18-21 year old who you served with, I can tell that it's you who is the one that hasn't grown up. That first look in the mirror when you take the uniform off for good can be a long and difficult one. It doesn't matter how well prepared you are. When you do something for so long, it literally becomes a part of you. 20 years ago, on October 28, 2002, I left East Chicago, Indiana for the unknown. I was 18 and had no clue what I was getting myself into. Before my 19th birthday, I was in a war zone, supporting Operation Iraqi Freedom. I would deploy three more times to The Persian Gulf, South China Sea and even be involved in piracy ops in the Indian Ocean, off the coast of Somalia. I had a small stint in Kuwait, which ended up being fun to say the least. At my 10 year mark, I willingly chose to go live on a small island in the middle of the Pacific Ocean called Guam. It would end up being the best thing that I ever did with my life. After that, I came back to California in 2014 to finish what I started. Now, the mission is complete.

Me and my brothers paid for all of that travel, school and good times with sleepless nights, late nights, blood, sweat, arguments, fights, fires, flooding, man overboards, toxic gas casualties, separation from our families and sometimes death. So anyone who says that we got all that travel for free, I say this with the most malicious intent and all disrespect, you can send y'all mothers to suck our military dicks. If they aren't available, send your wives or sisters. Hell, I smashed a few wives in my younger years, but that's a different story for a different time. For this statement, I don't apologize. Too many people automatically make assumptions about us, so these disrespectful words are free. I'm not under Uncle Sam's watch anymore. So, I'm letting the true feelings fly. Plus, I am, and will always be a sailor, with haze gray running through my veins. We cuss, drink alcohol, cuss some more, drink coffee, and cuss some more. Actually, here's the word FUCK for no other reason than to say it. Along this military journey, I lost all three of my grandmothers, my dad, my sister Michelle and Jackie Dunaway, who was an angel on earth who I simply called Moms, because she was literally like another mother to me. She wasn't my birth mother, but she gave me unconditional love, and that's important as ever when you are a teenager that's 2,500 miles away from everything you've ever known. I lost Connor, Walters, Achillez, Justin, Ralo, Double M-G Mark Mayo, Sterling, Ebony, Heather, Jesse and others I called Brother and Sister during these 20 years. Along with those losses, I had a lot of wins. I became a 10x published author (this is #11), a Top 25 world ranked poet (2019), a top tier marketer, all of that. I got to visit 27 countries on five continents. All in all, that was some hard shit to accomplish, when you consider all the obstacles that were in my way. For my friends that's reading this wondering. When we go into this shit, we find out that it is one of the most dangerous things we will ever do, whether Navy, Marine Corps, Air Force, Army or Coast Guard. War isn't even the scariest shit. Our guard is forever up, all eyes stay on us and we literally have no choice but to interact with our brethren, who are a bunch of ticking time bombs like me. There are so many different personalities to wrap your head around. We may not like everyone that we encounter, but that person you don't like may end up saving your ass one day. Most importantly, there are unwritten rules and guidelines to this shit. This book will not be filled with funny stories, or the typical party, sex and raunchyness that the military is known for (Especially the Navy. We literally wrote the book on acting an ass in different places around the world). This is the real life shit in and out of the uniform. There are probably, at any time, at least 10 military members out of 11 going through some mental health related shit. I say 10 out of 11 because there is always that one that just doesn't give two fucks about anything. Yet and still, we are expected to run a successful mission, all in the name of "freedom and democracy." This is truly an all hands effort. When we hear one dead, two dead, eight Sailors/Marines/Soldiers/Airmen/Coast Guardsmen missing, it pains us. Now, imagine the shit that we deal with when death or injury isn't involved. I took pride in this shit, even though I didn't let it consume me to the level where I thought I was better than anyone else. Or, the level where I felt I needed the military to be successful

(We'll speak more on that later). You won't see me stand at baseball games when they ask for military vets to rise and be acknowledged. You won't see me bitch if an establishment does not offer a military discount. Nah. I simply took pride knowing that every time my guys needed me, I never had to explain to a spouse that it was my fault that their husband or wife didn't come home. We may let the jokes fly, but when it's time to get down, we're with the shits. If only those above us (military and appointed civilian leadership) really gave a shit about what matters most. Mental health. Welcome to the real battle that occurs within the military. The one that nobody sees. The armed forces are already in a shit storm with commands lacking critical discussions on the state of race in America, particularly centering around their Black, Latino and minority service members. We had the remains found of the Army PFC at Fort Hood. Just a quick sidenote. I don't want to say her name, because it upsets me to talk about shit like this. This cover shit up, brush it under the rug mentality that the high brass has is beyond toxic. Fuck all the Army vs. Navy jokes, or any branch jokes for that matter. This shit is critical. If you can't protect your people, especially the Black, Brown and female service members, then you need not be in a leadership position. I don't care if you're the newly frocked E-5 or the senior O-6. People are fucking dying because of mental instability, caused by numerous factors that never get addressed. Women are being assaulted and can't get help. Men, are being sexually assaulted, and are afraid to seek help, due to embarrassment (I can count on two hands how many men I personally know that were sexually assaulted). Suicide is through the roof because asking for help for your mental health is frowned upon, and being tough at all times is expected. Black service members looking over their shoulder constantly cause the world is out to get us, inside and outside the gates. LGBTQ are feeling blackballed because of their sexual preference. Who you choose to worship still has negative effects. Immigrants who served are getting deported after their service is over. Meanwhile, this country fails to realize that everyone here is a fucking immigrant. Well, not everyone, but let me try to stay on course, cause I can go on all day about that shit. Members are being told that mission is more important than self and family. I did this for 20 years of my adult life with no time off, breaks or pauses in service. My active duty was consistent. I've literally had a tale of two careers. The first 12 were a worthwhile experience. The last eight, I saw my whole organization flip upside down. Now, more than ever, is the time for the military to focus on the mental health of their members, and not stupid shit like clean shaves. Oh, by the way. Navy, cut the dumb shit out about OSHA studies and will you get a seal in a SCBA (Self-Contained Breathing Apparatus). A former Gas Free Engineer who did over 150 drops, with a full beard, is writing this book. Clearly, I got a fucking seal. Now, imagine the junior Seaman or Fireman in the Navy. Hell, imagine the fresh, I haven't even been on the ship for a full week, individual, coming into this new Navy world. You have no duty, but you live aboard the ship and are relaxing. It's a Sunday and most of the crew is home. Hell, let's just say you reported on board two days ago, on a Friday. You've called your parents, excited as hell to feel like a Sailor. For shits and giggles, we'll just say they are from a small town, where experiences like this are the equivalent to winning best actor at The Oscars. You are chillin. Then, your first real welcome to the Navy moment is a big ass fire. I mean, a really big fucking fire. You are 18 years old, you don't know your foot from your asshole, and now you are in an FFE (Firefighting ensemble). You are trying to put the blaze out on a Sunday, when 90% of the crew is off work. This was the Bonhomme Richard in 2021. There are also a lot of other ships that we didn't hear about that have fires. They just managed to contain the casualty before it got out of hand. Trust me, casualties like fires happen all the time. Losing a whole ship? That's a different story. If it's anyone that we need to truly make sure is right, it's those E3 and below who had to witness and risk their lives during that shit. You may not think it, but stuff like that can be traumatizing. Work, family, life in general, these stressors wreck havoc on us. Especially, when the mission is stressed as more important than anything else going on with your life. Speaking of missions. I have always wondered. What indeed is the true mission for us to serve? That's a question that everyone will have their own unique answer to. For me, it is this book. This no shit, no filter, tell it how it is book, I see this was my purpose for serving. I hope in the end that this book opens eyes and saves lives before they are ever at risk. I'm not saying be soft. What I am saying is be mindful. Now, let's delve deep into this.

— 1 —

THE INITIAL MENTAL BREAKDOWN

I am a United States Sailor. I will support and defend the Constitution of the United States of America and I will obey the orders of those appointed over me. I represent the fighting spirit of the Navy and those who have gone before me to defend freedom and democracy around the world. I proudly serve my country's Navy combat team with Honor, Courage and Commitment. I am committed to excellence and the fair treatment of all.

From day one, when we entered the gates of Recruit Training Command, in Great Lakes, Illinois (About an hour north of Chicago), The Sailor's Creed was beaten into our heads. Now, as an 18 year old, fresh out of high school, I had no clue what this meant. The only shit that I understood was that I was a United States Sailor. Furthermore, I knew that I was expected to learn it and memorize it, along with the 11 General Orders of the Sentry, with the latter being the same across all military branches. Later, the only General Order I would truly ever care about was my 5th General Order. *"To quit my post only when properly relieved."* If you wanted to piss me off in a heartbeat, be late relieving me on watch. Now, let me break this down line by line. I am a United States Sailor. Well, duh. I enlisted into the United States Navy. I'm supposed to be a Sailor. When someone goes into the Army, they are deemed a Soldier. When someone goes into the Air Force, they are deemed an Airman. A Marine is just that. **A MARINE!!!** I emphasize this early in this book for my civilian counterparts because too often during my 20 years, I was hit with the "Thank you for your service, **SOLDIER**." Sheesh. I would think to myself, "Do they not see the words U.S. Navy stitched across my uniform?" I guess they didn't. More often than not, I just said thank you, and kept it moving. I know the Coast Guard is asking, well, what about us? Yes, you are known as Coast Guardsmen. Yes, your job is just as important. However, I was just keeping the focus on DOD, not DOT. Hey, it's still all love. I wished at times I could do the drug ops shit. The next part states, I will support and defend the Constitution of the United States of America and I will obey the orders of those appointed over me. That part right there, we have to get into that. It's no secret that I am Black. I mean, you saw the book cover, the Black fist, the back of the book with my picture. If you're blind, I apologize, and I hope that someone will convert this into an audio book. The history of Black people and the United States Government is not a good one. I could start with slavery, because we all know about that. That's been beaten into everyone's heads, Black or otherwise. Let's go into the era known as the Red Summer between 1917-1923, where dozens, even hundreds of all Black towns were burned, looted, and Black people were murdered in cold blood. The peak of Red Summer was 1919, but the highlight came with the government assisted bombing of the Greenwood District in Tulsa, Oklahoma, affectionately known as Black Wall Street (It was originally deemed Negro Wall Street by the great Booker T. Washington). I mean, if you wanna go back a few years prior, the President, Commander in Chief, Woodrow Wilson, held a private screening of "Birth of a Nation" in the White House (Really comprehend what you just read in that last sentence). Follow both of these acts with 'redlining' housing policies, segregation, urban renewal, and you can see why Black folks have a major distrust of the Federal Government. Trust me, the list goes on with police brutality, the wealth gap, a million other things, but let's continue. With all this said, how can anyone Black serve for a country that purposely seeks to destroy us? That is a great question, and I've been asked that a thousand times by friends, and sometimes strangers. Let me answer that. For many of us growing up in the quote on quote hood, we are looking for the first way out. Personally, I never wanted to join the military. It wasn't an aspiration of mine. In high school, I was primarily a B average student. Maintaining a 3.0 or higher was an easy day for me. I had both of my parents in the home, so they stayed on me about my grades, even though their own relationship was a F. Eventually, I did well enough to where one of my guidance counselors prepped something sweet for me. All I had to do was cut a 950 or higher on the SAT, and the prestigious FAMU (Florida

A&M University) would give me a full ride. Truthfully, I didn't care about it being an HBCU. Shit, my city was 30% Black. Chicago next door was 50% Black. Gary was damn near 100% Black, minus those White folks who stayed out near Miller Beach. Milwaukee, they were well over 50% too and only an hour and some change north of us. Trust, I grew up in the Black culture and saw Black everything all my life. I just wanted to get out of East Chicago, Indiana, my parents home and start my own journey. Well, come SAT time, I cut a 1020. I did it. I was happier than a high school offensive line getting treated to Golden Corral after winning a big game. I was about to get the hell out of here. Florida was on deck and I would see some more Melanin. Now, as I stated, I didn't particularly care that it was a HBCU because of the culture that surrounded me where I grew up. I had the Black experience for 18 years. However, I wasn't naive to the great people who had come from HBCUs. Plus, it was far as fuck from Indiana, and no sane man will chose to stay in Indiana over Florida, on any day of the week. Then, reality hit me. Reality being my parents. They told me they didn't have the money to spend on me for college. Wait a minute? I cut a 1020, earned a scholarship in architect engineering, all you have to do is get me down there, and you don't want to get me there? My goodness. Talk about a blow to a child's heart. With that, came the talks of the military. Now, I was seeing what this shit was really about. They didn't want to be involved in any way. More so, it was my dad. God rest his soul, and I love him, but it took years for me to get over that shit. Fast forward because I don't want to drag this story on, I went to the recruiters office.

The Air Force to be exact. I didn't want to be anything like my dad. He went to the Navy. I'm going to the Air Force. The Marines and Army didn't stand a chance in hell with me. I dealt with enough bullets living in East Chicago. I wasn't about to get purposely shot at like a dumb ass. As I walked out of the Air Force office one day, the Navy First Class who I came to know as BOATS, holla'd at me. "*I'll give you 10 G's to sign with me.*" It was some back door, cold cutta shit, but I would later learn that this was the way of the military. With that, I was game. The next day, I got my parents to sign the paper because I was still 17 and that was it. Now, in just a matter of time, I would be defending the same constitution that declared us as 3/5 of a person, with a 13th amendment that…well…I'll let you look that up. As I got to boot camp on October 29, 2002, the initial mental breakdown started. You were called every name in the book. You were placed with a bunch of people who you didn't know. Everything had to be done at a fast pace, so on and so forth. Those first three nights, you hated life and wanted to escape back to wherever you came from. We were turned into human pin cushions, receiving about 20 shots. What was in those needles? I had no fucking clue. Days later, they put a shot of penicillin in ya ass cheek. Literally, we were bent over a table, looking at one of our partners and they just stabbed your ass. Go ahead and laugh. Vets, you especially, I know you are laughing. I laugh at the shit myself now. After that first week, your RDCs (Recruit Division Commanders. Drill sergeants in other branches) ripped you a new asshole, tossed ya shit, called you every name except for Jesus Christ and made you feel like the scum of the earth. But, once you got the routine down the shit wasn't that bad. We were prepared for this for months in our Delayed Entry Program. Leaving home, getting yelled at, being told what to do. It was all routine. Now, looking back at it, this truly was the easiest part of my military service. Or, in the words of the famous spoken word artist R.J. Wright. "*Boot Camp was the funniest place that I wasn't allowed to laugh.*" I graduated on January 3, 2003 and was bused over to NTC (Naval Training Center) Great Lakes across the street, where I began welding school. It was six months of mischief. I was 18, around a bunch of other 18-21 year olds who had no guidance at all in life. $1,000 in my bank account was like $10 million to me. I went to school, played ball at the gym and took the train home on the weekends. Life was good, minus the occasional failed room inspection where I had to stand outside my door at attention, at 9:00 PM for a re-inspect. I was still in a comfort zone. I still had my folks at the crib an hour away. You couldn't tell me shit. Then, I represent the fighting spirit of the Navy and those who have gone before me to defend freedom and democracy around the world came to life. I was given my orders in late May. I was to report to the USS Nimitz based in San Diego, CA. I read the line on my orders saying *'The finest aircraft carrier in the fleet.'* It sounded good, until I saw that over the next 20 years, on every set of orders, it would say the same shit about whatever ship I was going to. I was excited, but disappointed at the same time. On my dream sheet, I put Hawaii, Spain and Florida, because you had an option to choose three. California was not even in my view. The only thing I knew about California were rappers and the weird shit I saw on TV that happened in San

Francisco. I truthfully didn't wanna go there at all. Florida was where I was supposed to be anyway, seeing how I was supposed to be at FAMU, watching the bands go dumb at halftime. How was I gonna make it out there? I emailed my sponsor as I sat in my parents house, counting down the days. When he finally hit me back, the breakdown occurred. They were in the Persian Gulf already, supporting Operation Iraqi Freedom. My senior year of high school, I watched the news in my second hour newspaper class of the Twin Towers getting blown up. I told myself that the military was out of the damn question. I wasn't about to be the next one to have a 21 gun salute. Yet, here I was about to fly to a war zone, with no clue of how to handle it. That fighting spirit came alive, and it was my first battle with mental health, even though I truly didn't know it. Imagine being an 18 year old being taken from everything you know, and flown half way across the world into unknown circumstances. The military didn't give us any class, lecture or any resources on how to prepare for such a drastic change. It was only, here's your orders and a few select words from those who had been there before. Most of the time, it was *"Suck it up," "You'll be alright,"* or my favorite phrase, *"You signed up for this shipmate."* I later learned that calling someone a shipmate was damn near equivalent to calling someone a bitch. The day came where I had to hug my parents goodbye. It was June 17th, 2003. I was still upset with them, but I had bigger fish to fry. I tried to envision what I was about to fly into. In my head, it was bullets, turbans and all out hostility. The first leg of my flight took me from Chicago to Norfolk, Virginia. There, I was picked up by a Petty Officer, who had told me to shut up calling him Petty Officer. What a hell of a way to make your introduction to an 18 year old who doesn't know any better right? There were a few of us heading over and we all piled into a van that took us to one of the bases out there to fly out some hours later. The next flight would be to Spain. As we took off into the dead of night, my brain was scrambled. I just knew I was a dead man walking. My time on earth was going to be short. I wasn't prepared for this. All we heard in boot camp and A school was "*You're going to a ship. Shit happens. It's real out there.*" Not at any point did anyone really break shit down. In higher up eyes, the military was simple. Since they had to learn the hard way, you had to learn the hard way. Basically, in a nut shell, it was fuck you, I did it. Suck it up, don't be a bitch and get it over with. After seven hours, I landed in Rota, Spain. It was daylight and early morning. I don't recall getting much sleep on the flight. When I walked off the plane, my lungs did a 180. I had never inhaled such good air in my life. For 18 years, all I knew was pollution from the steel mills that surrounded my hometown. I was actually breathing God's air now as I called it. Hold the fuck up? Was I actually in another country at 18 years old? Yes the fuck I was. For the next few hours, all my anxiety disappeared, as I hung with some older cats and got a little taste of Spain. Come the late evening, we were back on the plane, headed for Sicily. After two and a half hours in the air, we landed on this beautiful island. We dropped another batch of people off, had a few hours of downtime, and we were off again. This time, it was Bahrain. The Middle East. We were headed into hell as far as I knew. It was the shortest and longest flight of my damn life. We landed at 9:00 PM on the dot. I'll never forget that shit. Someone came on the plane, escorted us off and into a building. There were six of us, with at least three of us being newbies to this shit. We were given the rules on how to conduct ourselves at the hotel where we would be staying for the night. In the morning, we would be flying out to the ship, getting ready to continue in the tradition of those who had gone before us. In my room that night, I ensured to keep as many lights out as possible. I was scared shitless. I was alone, away from America, knew no one and I had the biggest fear that someone was gonna come in my room, kidnap and torture me. As I walked in the bathroom to shower, I saw that there was a window in the shower. I had to turn on the light to see, which was a bummer to me. I bathed in the back of the shower. When I had to come up and rinse myself off, I would duck, squat and let the water hit me. I didn't know who had what weapons aimed at the window, so I kept myself out of view. When I finished, I walked back to the bed with nothing but the television illuminating the room. I dried off, threw on my hoop shorts and curled under the covers. No one came to check on us. We were told to be downstairs at 9:00 AM to be bussed over to the terminal to catch a COD (Carrier Onboard Delivery) onto the ship. Now, I proudly serve my country's Navy combat team with Honor, Courage and Commitment. What was I serving and how could I prepare for it, if I didn't know the end result? What's the honor in that? What courage will I need and for what situations? Who was I to be committed too? All these questions and no one had given me a clear cut answer in my short almost eight months of service. The next morning, I stopped for breakfast in the hotel restaurant, near the lobby. I kept it simple with an omelet. I was tired as shit, because I didn't sleep well during the night. I had too many nervous thoughts going through my head. I finished up and about 30 minutes later, it was time. My new life had officially begun. We were packed like sardines into this small, baby plane. It was damn near like some shit out of Toy Story. There were six of us. There weren't any windows on the plane, so seeing anything wasn't even an option, minus each other. We took off and it rattled in the air like a modern day Spirit Airlines flight. That wasn't a good thing either. I had been in packed cars before with my friends, but this shit here was not the business. As I recall, we probably were in the air for 30-40 minutes. I felt when we descended to land on the ship. As I felt the roll of the wheels, and with God as my witness I am not making this next part up, we kept going. *"We missed,"* the pilot said. What the fuck did he miss? I wasn't even a legal age to drink and I know that I needed some crown on the rocks right about now. I knew we weren't crashing because I felt the plane go back up and make a hard turn. A few minutes later, we tried again. Success. At this point, I was ready to ask them for my discharge. This was my welcome to the Navy moment. A botched landing because we missed the arresting gear that helps you land on the flight deck. I don't know whose fault it was and I didn't care. All I knew was that I was already over this Navy shit. Minutes later, I would be in a room, with this huge ass manila folder, with all my stuff. This tall, skinny guy came and got me, took me downstairs and we started the process in his office. On the way there, I received mugs from men and women. The women were looking to see who the new guy was. The men were looking to see who the new competition was and to size me up. I would learn quickly that this is how the Navy operated, especially on deployment. As I got taken down to my shop, there was instant gratification. A Black man was running the shop. In school, There were only five of us in total. Through all the slick ass, racial undertone chatter that we had to deal with while there, we learned that HT, the Navy acronym for welders and plumbers, was a quote on quote White boys rate. To see a Black man in charge, that was refreshing. I got settled in and they found me a rack (the name for our beds). I put on my coveralls and walked back to the shop. Then, the shift happened. We were mustering up before dinner. All these brothers were looking at me. I was wondering who in the good fuck are these guys? Come to find out, they were all HTs. A full 10 of em. You talk about being brought back down to earth. I was running the earth right now. At this point, some people are wondering why in the good hell am I highlighting Black welders. You may be part of the ALL LIVES MATTER crew. Well, let me educate you. In 20 years of serving, I may have served with a total of 70 Black HTs. Some of you are thinking, well, that's a lot. Consider that my first ship had well over 100 welders come through in my three years onboard. Again, we are not looked on as welders. When you're Black and in the Navy, most assume you must work in Supply, work in IT or you are an undesignated person without a job, so you belong to the deck department, who takes all the jobless people and basically has them working as sea going janitors. In the Navy's beginning days, most Black sailors were designated to the Supply Department, which were the cooks, folks who washed and pressed clothes, and stocked shelves. What does that sound like? I don't even have to explain that. To see this barrier broken down early in my career, and know that I would have people that looked like me in my work space, that was mental score number one. I still battled with the separation, being thrust into an environment with no explanation other than, it's your job. However, I felt a little bit at ease knowing that someone who I was working with could understand me and shit that I was going through. This leads me into the last line of The Sailor's Creed. I am committed to excellence and the fair treatment of all. In my eyes, I had Black excellence around me, and I just know with these guys here, fair treatment was gonna be a no-brainer in this military world. I was wrong by a long shot.

— 2 —

NO ONE IS HERE TO SAVE YOU

And I was alone, had been for a while, and might be for a while, but it no longer frightened me the way it had. I was discovering something terrifyingly simple: there was absolutely nothing I could do about it.

James Baldwin, Just Above My Head.

November 5, 2003 came around faster than I could imagine. Over the last six months, many things had occurred. For one, I was a year older. I was still a baby, but I felt like a grown baby. I had completed my first deployment, I didn't get my shit pushed in by the old salts and I made it to Spain, Italy, Bahrain, Dubai, Singapore and Hawaii. Five countries and one exotic state visited by the age of 19 was one hell of an accomplishment. I can hear some people now saying *"Oh it doesn't count cause your job got you there."* Same way your mom's one time refusal to swallow got you here. So, we're even. What they also don't take into consideration are the long nights of watch standing. The getting up in the wee hours of the morning because a chem light has dropped, or been thrown into the water, and now, man overboard is called because there just may be someone in the water. There are casualties that happen out at sea. Fires, flooding and toxic gas. Unlike everyone on solid ground, we can't dial 9-1-1 and have the fire department come and save us. We have to do it ourselves. Add on medical emergencies and 14+ hour work days, and we deserve every bit of fun we get entitled to. So, again, to anyone who deems that we are just catching a free ride on a warship, I have two suggestions. One, sign your name on the dotted line and do what we do. Two, just shut your fucking mouth before a vet punches you in it. I'm sorry, but I had to put that out there. Now, back to the story. November 5th was the day we came home. I had bonded with a lot of people and became an integrated part of the crew. I learned that trust was something that I had to learn on the fly. You always found out who you could rely on when shit hit the fan. I looked out onto the pier at Naval Air Station North Island (NASNI) as families waited with signs for their loved ones to come off the ship and greet them. It was a site that I'd never forget, but also one that would turn into something that I wish I could forget. I trickled off the boat with the rest of the crew, looking at hugs being handed out, kids running up to their parents, all that. I went across the street and sat on a bench, under a huge tree. This was my first time in California and the sun was out. It was a beautiful day. Not bad for my initial introduction. I watched one by one over the next two hours as the crowd dwindled from thousands of people, to just one. Me. There I was, in my dress blue uniform, lonely. I had no family waiting on me. I had no phone or car. In reality, I had no friends. The cats on the boat were cool, but I only knew them because we had no choice but to get to know each other. Now, we were home and I was a cactus in the middle of a dry desert. I w noooo aited a while before I went back on the ship to get ready and do my own thing. The berthing was an eerie quiet as everyone was gone. I put on some of my civilian clothes, left and began to walk the base. I had no clue on where I was going. I just knew that I needed to find a place to eat and somewhere to call my parents to let them know that I was alive and well. I eventually found both. The single sailor center was directly across from the bowling alley. I was able to use a phone there and call my folks. I tried to tell them as much as possible about my adventures and plan my trip home. My plan was to come home on Christmas so that I could feel a sense of normalcy again. In the meantime though, I needed to organize a plan to get my shit together. Again, I had no phone, car or anything. I couldn't bog anyone down on the ship cause they all had obligations. Luckily, my dad would get in contact with a retired Marine cousin of mine and I could start the process. Until then, however, I would just have to ride it out. I left the center and went to the bowling alley. Wings and fries were the best thing for me right now, along with hitting six or seven pins at a time. I wasn't a bowling expert. Six or seven was my lucky number. If I got a strike, I felt like a man who had put the toilet seat down for his wife and considered that a major achievement. I needed my MVP trophy. The fun lasted for an hour and some change. Then, I found myself bored of this. By the time I left the bowling alley, it was dark out. Base was a mere shell of itself, as people were home on this Wednesday night, having dinner and tucking their kids into bed. Me, I was a lonely 19 year old who knew no one, walking back to a place that I had just spent the last six months living in. I know that we are all unique and should enjoy being who we are. In this moment, however, I wish I could be one of those people, relaxing on a couch, watching television. No one prepared me for this. Truthfully, no one could. Hell, this is the life story of many new faces in the military. The longer I served, the more I would come in contact with this type of loneliness. Except, in my senior years, I had to be the parent figure, or friend of that young male or female sailor. I remember one of the most disturbing instances I had to encounter with new people. It was a Tuesday night in early 2020. I was stationed at Camp Pendleton, at Assault Craft Unit 5. It was a normal duty night, and we had a new guy coming in. I volunteered to go pick him up. Anyone who has served knows that anytime you can get away from the command, using government gas, you take it. I made the 45 minute drive to San Diego Airport. I hit the USO before anything and got a quick bite to eat. If they didn't have anything else to eat, they were always good for a sandwich and some chili. Sho nuff, that night, it was chili on deck. I told 'em to put that shit over the two hot dogs I got and I was gucci. I looked at all the Marine recruits waiting to come get picked up and cracked a smile. Good luck is all I could think as I nodded to the DI's who were there. I finished up my meal and went over to baggage claim, waiting for the kid. After 20 minutes, he was headed down the escalator.

He wasn't hard to spot. He had on a backpack and had that new guy look, with the I don't know where the hell I am walking demeanor, with a terrible haircut. I mean, it looked like someone had cut his shit with some bear claws. Yup, he was definitely a booter. I introduced myself and he kept calling me Petty Officer. It was annoying, but I flashed back to that day in Norfolk, Virginia, when I was a young guy who was just trying to show respect. There was no way I was gonna treat this kid the way I was treated. With a simple, calm tone, I simply told him hey, you don't have to call me Petty Officer repeatedly. Just say HT1. He responded with yes Petty Officer, and I just let it go. Deep down, it was funny. He would literally have to have boot camp shaken out of his system. Mentally, I had to remember where I was in my career and where he was. I kept saying Norfolk, Norfolk, Norfolk in my head. Mental therapy victory. I didn't belittle the kid, or make him feel like shit for showing respect. We got his bags and I took him over to the USO to grab a free bite to eat. As we entered, I told him to grab what he wants, drop a deuce, take a pis, whatever else. It was already past eight, damn near nine and I was good for the night. Once he settled on a few sandwiches, cookies and a pop (not soda for you uncultured folks), we made it back up front. As I was about to head out the door, I overheard the conversation occurring at the desk. There were two sailors in their dress blues, one ribbon only and they sounded distressed. Yea, they were definitely fresh out of boot camp. They were asking for help from the staff, who were trying their best to assist them. "*Where are you going, sailor?,*" I asked him, ensuring that the tone was calm, nurturing and one of I am here to help you, and not that I'm a big, badass First Class Petty Officer who has pissed more sea water than you ever sailed. He told me that he and his fellow sailor were reporting to Point Loma ASW(Anti-Submarine Warfare) base, but no one was here. They had called repeatedly the numbers they were given, but no one showed. Mental health breakdown had occurred once again. It was one thing for me to report to my ship with a bunch of unknown people. However, at least I knew it was a bunch of us on a plane, all going to a destination where we would have a bed, food and people around. Plus, I made contact with my sponsor before I got there. These kids had literally flown into an unknown city and received no help. In the Navy we call this the first 72. The first 72 hours can make or break a new person. They will either feel like shit, or feel like they will be the shit. In this case, they were both already in the toilet and just waiting to be flushed. "I got 'em," I told the ladies at the front. We walked to the van as I made small talk with them, to put their minds at ease somewhat. This was a bad look for whatever command they were going too. I made the quick five minute drive to ASW base and headed towards the main building. I didn't know in particular what command they were checking into. Honestly, I didn't know too many commands on this base, as I was more familiar with the Point Loma Sub Base. I parked and helped them get their things out of the van. As we entered, I saw three of my fellow First Classes at the desk, which added some ease to this situation. I could deal with people at my level and not a disgruntled khaki whose kids probably didn't like him. I rapped to the girl at the desk, telling her that I found these two at the airport and they had no one there to pick them up. She took their orders and then said some shit that appalled me. *"No one told us anyone was coming in tonight."* Wow. Really? This is how we are treating sailors now? There are three people on watch and they have to scramble around to find out what command dropped the ball. This was sad and pathetic. Then, the Navy wonders why today's kids have a lot of no fucks to give. When they need the Navy to give a fuck, they don't. I left back headed towards Camp Pendleton a little pissed off. We ask so much of these kids today, but we can't do simple shit like pick them up and be welcoming when they are arriving in a place they have never been to. Welcome to mental strain 101. No one gives a fuck about me. So, why should I give a fuck? It's a great question and it's one that the military as a whole still hasn't answered.

I've discussed the trauma and mental strain of coming home from deployment with no one waiting for you. I've told you about these kids who had no one there for them. Loneliness, however, gets more dramatic than that with this organization. I just need you all to stay with me. I'm going to bounce around with a lot of stories from different time periods. Just know, it's all going to flow. You'll truly understand why I wrote this short, yet efficient story in the end. The next ounce of loneliness involves a friend. A dear friend. I'm three years into my career now. It's 2005, and I'm on my second deployment. I'm not a veteran in the game, but I played in it long enough at this point to know the basic rules. Be where you need to be on time. Do your job. Don't cut corners. If you have a question, just ask. Don't ever assume. Making assumptions in the fleet will get you hurt or killed. Or, it will get someone else hurt or killed. If you follow these simple rules, you would be good 99.9% of the time. Oh, I almost forgot. Don't snitch on your partner in a foreign port. We don't do that shit. I was now a Third Class Petty Officer. An E-4 for you non Navy folks. I wasn't making any life changing money, but the tax free status in the Gulf sure made it seem like it. Truthfully, this deployment saved my life. When I received my $10K enlistment bonus in 2003, I had no clear cut sign on how to invest it. No one told me anything meaningful except the usual save your money. Well, except for my big brother Reggie. He told me to be smart with the shit. Hell, without him, I wouldn't have a car, because he co-signed on it. Yet, what do you expect a 19 year old to do with a large sum of money that he has never seen before, with no guidance? Yea, I blew it. Throwback jerseys, more clothes, shoes, the club, more throwback jerseys, just stupid shit. You couldn't tell me anything because I figured since I had a car, phone and my cousin's house to crash at, I was good. When May 6, 2005 came around, I saw I had less than $100 in my bank account. At that moment, I told myself that I would never be in this predicament again. Deployment went off with the usual shit. We stopped in Hawaii, which was always the last wind up before shit took off. Then, the game changed. We went to Hong Kong. From there, we headed to South Korea, but North Korea was trippin, so we shifted down to Guam. Remember Guam. Trust me. More about that place later in the book. From there, Singapore, then Malaysia. My mental health was one hundred miles an hour and running. All this partying within a two month span gave a man new life. Not to mention that Asia is dirt cheap, so I didn't have to spend a lot to have a good time. I was truly living the good life. I was so used to standing watch and long nights at this point, that it didn't bother me anymore. Once the two month fun run was over, it was back to the Straits of Hormuz. Two years ago, I was scared shitless entering into The Persian Gulf. Now, it was nothing. We got into The Gulf and pulled into the usual. Bahrain and Dubai. Life was crazy at this point. Mel Gibson was filming us the entire deployment for a documentary called CARRIER, so there was something new and exciting everyday. Some of that exciting shit was good, and some of that exciting shit was definitely raunchy. I'll never tell our secrets, but boy oh boy. Some of the shit that went down was good enough for a full release movie. Truth be told. If he released the uncut version, with everything the camera crew caught, all 6,100 of our crazy asses would have had a bad conduct discharge. However, what I am about to disclose wasn't a movie. This was real life. I headed to the berthing one August day in the middle of deployment. I don't exactly remember what I had to get, or whether I was just heading down for a quick break. As I made my way into the aisle, I saw a co-worker. A fellow HT. He was packing up his stuff and balling tears.

"You good man?," I asked him.

"My son just died bro."

At that moment, my heart sank. The breath left my body. I didn't want to comprehend what he had just said. Speechless, I did the only thing I could do, and that was to give him a hug, as his tears soaked my coveralls. We weren't best friends, but we were good acquaintances. I don't know how long I hugged him for, but in that moment, comfort was all I could give. His mentality was one of sadness, loneliness, emptiness. My mind was destroyed. I was only three years deep, but I knew that if I stuck this out for the long haul, that at some point, I would have to deal with the grim reaper, in this exact same place, feeling the exact same way that my guy was feeling. Helpless and alone. As he prepped to fly off the ship and back stateside, I was in a whirlwind. Thoughts of death started to cross through my head. What if I got the call that my mom, dad, sister, cousin or someone close I know had died? That is some shit that I didn't comprehend until that very moment. I was young. 20 years old now to be exact. I was still in the mode that I had time. The thing was that I was young, but everyone else wasn't. It caused me to gain a whole new perspective on things. For the rest of deployment, I was nerve wrecked every morning at quarters. That *"McClain. Can we see you in the office?,"* question was just a ticking time bomb waiting to go off. Who would it be? I don't know. The only friend I had lost up to this point was an electrician that I worked with. Cool ass Filipino dude. He was coming back from the casino one night, when the vehicle he was in got involved in a collision of some sorts. I'm not sure to this day whether they were hit, someone lost control or what not. All I know is that he became the first casualty I ever experienced while I was in. Death had now taken on a new meaning. I grew up in the funeral home with my dad, as he worked there for over 30 years as a primary mortician. He drove the hearse at times, filled in as a pallbearer, but his bread and butter was working on those bodies. This was something different. It felt different. I couldn't quite put my finger on it. However, as with us all, it wouldn't be the final time that I sat with death and had a drink with him. We finished out the rest of that deployment without a hitch. Most in the Navy call it WestPac. I called it Crunk Pac, because it was live for the entire six months. Mentally, I had peeled every layer off. Happy, sad, lonely. For someone else, empathy, all that. It was building me up for a more complex world that I still didn't know existed. Hell hath no fury like mental health that didn't get its regular dose of much needed therapeutic medicine.

— 3 —

MAMA TOLD ME THERE'D BE DAYS LIKE THIS

It's not about being in the right place at the right time;
it's about being the right person, even if you find yourself
in the wrong circumstances.
Mark Batterson

When we hear the day September 11th, a lot of emotion comes into play. I talked earlier on how I was a senior in high school when it occurred. Now, just imagine being a member of the armed forces when this day occurred. I wasn't enlisted then, but I heard all the stories. How the bases went to Threatcon Delta, which was the highest level of force protection. Ships were pulling out and missions were changed. Basically, shit had hit the fan and it got real. By the time I came in, there was war in Afghanistan and Iraq. However, as we all know, the biggest war anyone will ever have is with self. By June 2006, I was done with my first ship and at my next command. This would be my first of two times at Camp Pendleton. Here, I was now a 21, soon to be a 22 year old young man. I was grown, but I hadn't grown up. That would change quickly when the biggest shift of my life occurred. On the Nimitz, I was surrounded by people who made me comfortable being me. Here, I was surrounded by men who were gonna give it to me straight and force me to elevate out of my personal element. Smoke, Dre, Mo, Jack, Donnell and Eric Finch. Yes, that Eric Finch. The one everyone now knows as the celebrity Snoop Dogg look alike. Y'all know the celebrity. I know Eric, one of the best mentors anyone could ever have. We are not the same. I'll describe each of these individuals in depth. Eric was originally from Louisiana, but like me, found California as where he truly belonged. He was funny as life. Jokes were an understatement. This dude was the life of the party. He had the roasting skills of DL Hughley. The fire off style of Bernie Mac. The no fucks given aura that Eddie Griffin possessed. I mean, you put him in a room with 10 people and by the end of the night, there would be 10 deaths, as they all would die from laughter. He was and still is a big OG Unc. Donnell, he was straight OG. RPs in the Navy are in a nutshell, the Chaplains assistant. Don't let that shit fool you. Most of the RPs that I know are some wild dudes. They will literally start and end the party, whether it be in a good way, or in an air it out way. Donnell, he was truly an OG from the blocks. Born and raised in LA, he ran the streets, doing what he had to do to survive, much like a lot of men during the 70s and 80s. You can tell in a man's eyes when he is telling the truth. There was no fake shit in his bones, voice or demeanor. He had a lot of knowledge, and looking back, it was wise to soak it up, as I still do today from him. MO was another LA bred cat and he was with the shits. What's understood doesn't need to be explained, and some things that happened don't need to be shared with everyone. Just know he had my respect and I wasn't going to test his gangster at any time. I didn't want to meet that side of MO. The one that was Crippin before the Navy. Like, certified. So, before any of y'all think some men can't bring the streets with them into another world, you are sadly mistaken. As I said, some stories aren't meant to be told, but trust, he wanted all the smoke on or off base. Smoke was a bald headed, laid back, funny ass dude. He reminded me of the uncle at the bbq that was sheer entertainment. If your girl was ugly, he'd be the Uncle to say lemme talk to you over here. If she was bad, he would give you the International Black Uncle approval line of *"Aight nephew. I see you."* Jack. Well, Jack was a lot of things. If there's one thing I learned from him, it was to drink a beer when you first enter the club, and put that bottle in your back pocket. I was never the type to fight in the club. I went there to have a good time. However, one night, the methods he taught me came full circle. I was in the now defunct Blue Haven in San Diego. Let's just say the problem ended as quickly as it started and I finished my night without a scratch. Then, there was Dre from East St. Louis, Illinois. That told me all that I needed to know. Over the years, he would come to be one of the primary reasons I am where I am in life today. It was more than just the Navy with him. He was trying to prepare me for something greater. Sometimes, it took tough love. That's just the way it was. Part of that love would be shown on September 11,

2006. Aboard Camp Pendleton, they had a no shit tackle football league. This excited me to the core. I had last played football in high school, so to put on the pads again was like winning a million dollars to me. The command football team wasn't deep in numbers, but we had enough to go out there and compete. I was geeked up. About two hours before game time, I arrived at the field we were playing at that day. Out of all the fields on base, we were at shits creek as I call it. *"Man, someone's gonna get hurt on this joint,"* I told our QB, who was out there with me. In The Bible, they say life and death is in the power of the tongue. I was about to find out. Second quarter came around. We were down 7-0. I lined up in the slot. I was a defensive player all my life, but I was gonna go both ways with this. This was my release therapy. Hike! I ran my five and out route. The ball came. I caught it. I turned up the field, got tackled from behind and POP!!! The nastiest sound I had ever heard rang through my ears. I instantly dropped, rolling around and eventually tossing my helmet. I knew it was bad. I felt it when it popped. Just from me watching sports all my life, I knew my ACL was gone. I remember Coach coming out trying to keep me calm. I managed through the pain, but it was some tough shit to deal with. Dre and Jack eventually carried me off the field (literally) where the ambulance was waiting for me. Remember I told you that I was stationed with men who were gonna tell it straight? MO didn't let that ambulance take me. He had me put in the whip with him, and he drove me to the hospital. How does this affect mental health you ask? Easy. It was the first real lesson I had on being a man. When it's man down, nothing else matters until your man is up. It's the same in the fleet. When shit hits the fan, you have a choice of two. Stand up like a man, or run from danger. Out there, on the water, running from danger while your guys are fighting just so you can stay alive can have dire consequences. Once all is said and done, you were gonna have to answer to them. Here, in this moment, I was the man down. A young 22 year old who was injured, and completely depressed. OG's who I had only known for a little over three months were there to pick me up. I sat in the hospital that night, waiting for the results of my scan. I knew it was bad, just not how bad. The great thing about hospitals on Marine bases was that they had great service. You come in, and you don't have to wait long for anything. On the Navy base, you can go in for a broken arm at 6 a.m. and you were lucky if you were out of there by 10 p.m. Navy doctors had more urgency on Marine bases because they dealt with a lot more traumatic cases being around them. The nurse came out and wheeled me into the room. *"You see all this right here? Yea. That's supposed to be white. It's all black. Yo shit gone dawg."* He was a brother, so he told me in only a way another Black man can understand. My ACL was torn, along with my PCL, lateral meniscus and MCL. In a nutshell, I was fucked. I pulled the Daunte Culpepper. At this point, football became the last thing on my mind. The first thing was that my pops was coming in next week, so I had to tell him. He was gonna have to get a rental and drive me around. That was issue number one. Two. I thought long and hard on this. Was my Navy career over? I was on sea duty, which means I was deployable. If I couldn't deploy, then that means I couldn't fulfill my purpose with this organization. The Navy kicked folks out for less, so this was weighing heavy on me. I had two years left on my contract, and I was in no way prepared for life if they gave me the boot. I tried to keep my nerves calm, but it's difficult when you haven't even reached the pinnacle of your life and everything feels as if it was snatched from you. Two months later, after intense therapy to attempt to achieve some range of motion, Navy surgeons performed reconstructive surgery. Now, the real battle had begun. Me vs. me. I knew for at least the next nine months, my will would be tested. I had to heal physically, still go to work on a reduced schedule (in a different role) and drastically change my lifestyle for the next few months. As with any young person, I was hard headed and didn't listen. I was in the club on crutches. I was forcing myself into situations I knew I couldn't escape if I wanted too. In my head, I was gonna get kicked out eventually, so I might as well live it up while I still had a steady paycheck. I partied, I drank, all while fighting in therapy three times a week to get back to full strength. This shit was hard. Actually, I wanted to just quit at times. I couldn't for two reasons. One, my guy Kevin who worked with me, sacrificed his time and energy to take me to therapy and get me out, so giving up on myself would be like giving up on him. Secondly, I was in physical therapy with a Marine named T. He was from North Carolina. Everyday, he kept a smile on all the therapists faces. That in turn kept one on mine. When I was back to driving, I would head over to his crib and play him in Madden. 95% of the time, he was taxing me up. Not gonna lie. He became one of the few people that I couldn't beat except for every blue moon. All in all, it kept me sane, along with the drinking and clubbing. May 2007 came around, and for the first time since September of last year, I did a full sprint. I damn near cried because I was so happy. With that, however, I knew it would be the time to become fit for full duty. That was cool, until I got hit with the news. MedBoard. In a nutshell, it's a board conducted to see if you are fit to continue military service after an injury. I had one of the worst knee injuries, and I felt fine. However, would the military deem me fine and give me the ok for continued service? At this point, I had under a year and a half left. Just let me finish my time and roll out, is what I thought. Once the findings came back, I was cool. I was found fit for full duty, good to stay in and I went back to work. Hell, they even let me start playing football again. It was now August 2007. This time, I played with a different squad. The 7th E.S.B. (Engineering Service Battalion) Wildcats. It was a comeback for the ages. 74 tackles, 11 sacks, 4 forced fumbles and 2 fumble recoveries. I finished as one of the top defensive players in the Camp Pendleton Football League, and also made All Camp Pendleton. What stopped me from playing further? Deployment. It was about to be my third one in six years, and at this point, I had enough of the United States Navy and them taking my summers away. As higher ups let me know, either I go on deployment, or they separate me early in April 2008, instead of October. With that, I had no problem, until I really thought about it. Was I really gonna subject myself to going back to Indiana? East Chicago at that. I loved my city. I will always love it. However, it wasn't conducive to what I wanted to do in life. The last almost six years had seen me travel the world. I was also a part of some cool shit that only the Navy could provide you with. In 2005, The University of Oklahoma and The University of Oregon were playing each other in the Holiday Bowl, and the teams were having lunch on my ship. As with everyone else, I wanted to meet Adrian Peterson. Yes, All Day AP, who is a future NFL Hall of Famer. Cool shit like that was amazing. Mel Gibson and crew filmed my entire second deployment, and Mekhi Phifer, Gale Sayers and a shit ton of more people all came out. It was things like that which made me smile and appreciate it. It was the petty bullshit that I was tired of. Was I tired of it that much to return to the same place I wanted to escape for 18 years? Nah. I wasn't. So, to the tune of $15,000, I re-enlisted for another 3 years, even though I couldn't sign the contract until October 1st. The five mental plagues had hit me in a row regarding my life and career. Who was I going to be after my injury? Personally and professionally. What was it that I really wanted to do with my life, because I still didn't have a set path, and the last two years only showed me that I still had a lot of growing up to do. Where was I destined to be? California seemed like heaven compared to where I came from, and the last thing I wanted to do was to go back home and work in a steel mill. I wasn't too good for any job, but I for sure knew that I was way more talented than to be making steel, the same thing my dad did coming up. When was I gonna have my reality check moment? It was hard to ponder that the worst in life hadn't yet occurred. Lastly, Why was I here? Meaning, in this position, in this physical place, in this mental space. My mind was diluted by a false healing with alcohol the last two years. I knew that temporary cure wouldn't be efficient anymore. Funny shit is that the answer to my WHEN question, would come sooner than I ever imagined.

— 4 —

WHEN YOUR HEART STOPS

October 25th, 2008 wasn't any other day out to sea. It was Saturday, and on Saturday's, we didn't do shit. The weekend on deployment was similar to civilians in their work space. In our case, it was more of the wind down. We usually had the fun shit on Saturday night. It would start with chicken wings and pizza for dinner. I know what y'all civilians are thinking, but trust me, this wasn't anything special. Anyone who has served in the Navy knows that the menu doesn't change but every blue moon. Let me give you the rundown on how we do shit when it comes to the menu. Monday was usually Filipino day. This was when we had lumpia, chicken adobo, pancit, shit like that. Plus, you know we had rice. If Filipinos love anything in this world, it is rice (That's not a stereotype. That's a fact. I saw it for 20 years. Hell, they will tell you). Rice to them is what a good batch of fried chicken is to Black folks. If they switched the menu up, we might have hamsters. No, we weren't eating the little furry creatures that ran on a baby wheel. This was our name for chicken cordon bleu. Tuesdays were Taco Tuesdays. Tacos, fajitas, enchiladas, quesadillas, depending on what ship you were on. The sides were corn, refried beans, shit like that. Wednesdays were burger and fries day. Oh, don't think you were getting some gourmet shit. Now, maybe on the big decks, you'd get the freshly made patty, with the bacon that was cooked there on the grill. If you had some bomb CS's (Culinary Specialist) like I had on the USS Nimitz, USS Bunker Hill and USS Harpers Ferry, you'd get that sit seasoned up and all. For the most part, you got that frozen shit beef that came in a white box. If it's one thing the Navy did, it was provide the cheapest food available. You had to have some extraordinary cooks that could turn shit into sugar to make some things taste good. You really have to live and experience this shit to get it. To go along with burgers and fries, sometimes corn dogs were added. The sides were usually tater tots, some potato wedges or maybe some mixed vegetables from the can. Thursdays were Italian days. Spaghetti and some trash ass chicken parmesan. Let me tell you. All five ships that I had the honor to serve on did chicken parmesan the same way. A breaded piece of pre cooked, processed chicken that was warmed up in the oven. From there, they'd throw some pasta sauce and cheese over that joint. They put it in the oven for a bit more and boom, there it was. It was more like Shitty Parmesan than Chicken Parmesan. Even the greatest CS in the fleet couldn't fix that shit if they had Guy Fieri's hands. Friday was whatever. That was kind of the fuck it day. I mean whatever the fuck they pulled out of the freezer, that's what they were cooking. Now, don't get it twisted. CS's catch a lot of flack. Many talk about them like dogs.

They don't cook this or that. I know you've seen the viral facebook pics where people would put two baby wings on a plate with a piece of broccoli, talking about "This is how we eat." I can tell you that's a whole damn lie, made up by this new social media generation of kids who tryna go viral, thinking they gone get some pussy off a post. They are some lame shits to me. Ain't nann day in 20 years I ever got just two baby ass wings and a slice of bread. For the majority of my career, I had great cooks. I ate well. A lot of good folks I met in the Navy were cooks. They may not have been Emeril Lagasse, but they damn sure weren't burning the joint down. So, shout out to all my cooks. I got love for ya. Now, back to Saturdays. As I was saying. We would have chicken wings and ice cream on Saturday nights. With that usually came something fun like karaoke, Uno, spades or domino tournaments. Hell, even if it wasn't a tournament, we were getting it in with those cards. This October 25th day signified that we had 10 days left until we got home. This was gonna finish out my third deployment and man was I happier than a fly in a pile of shit. The travel was cool, the partying was wild and the foreign women were a whole other story. However, there was no better feeling than knowing you were about to head back to the states. That night, I partook in our weekly poetry night that occured in the ship's library. What started as a small gathering, created by one of my partners from Dallas, Texas, turned into a big event. Looking back on those days, I sucked. I was terrible. I know some of you who know me thinking, naw fam. You were cold with it. Yea, about as cold as a lizard's ass in the middle of the desert. I wasn't always a monster with the words like I am now (professional spoken word artist). However, back then, I thought I was the shit. I finished that up and headed down to the barbershop. I already knew what was going down in that joint. Dominoes, shit talking, cards, Madden, and more shit talking. Leading the pack was Master Gunns Brown. He was the Triple OG of the ship. He was a well respected and decorated Marine who treated people fair, but you didn't wanna cross him. He was funny as shit, but he'd bust ya ass quick if he had too. He was a dark, bald headed brother from the boot. Louisiana for y'all that don't know. He was on the domino table with all the boys, talking his talk. Tonight, however, the talk wasn't all about dominoes. We were down there having a major debate. Who had the better release that year? Lyfe Jennings with *"Change"* or The Dream with *"LoveHate"* album. I mean we were in a full blown argument at times. We listened to both albums straight that night and critiqued every song. **"LYFE!" "DREAM!" "YOU CRAZY AS HELL MUTHAFUCKA!"** That's basically what I heard for almost four straight hours. By the time three in the morning came, I was still upbeat. This was normal for me and so many of the younger cats on the ship. We could play bones til two or three in the morning, get three or four hours of sleep, and wake up to do our job like nothing happened. I was 24 as well, so my body's check engine light hadn't come on yet. I eventually called it a night and went back to the berthing. I took a quick shower cause I don't go to sleep with my ass smelling funky. I slid into my bottom rack and dozed off effortlessly. Tonight was a great night and in two days, I'd be in Hawaii for the God knows millionth time. Sometime after 8:30

AM, I got up to take a pis. When I slid out my rack and stood up, my heart was racing. I ain't no what the hell was going on. I know I didn't have a nightmare or wake up too fast, so why was my cardiac going like I drank 10 Red Bulls? I had no idea. I shook it off, got my dental hygiene joints and went off into the bathroom. It was cool here. I brushed my teeth like normal and headed back into the berthing. Then, it happened again. That racing feeling. I took some deep breaths, trying to slow everything down. I didn't know what the hell was going on. Aight Joe, I told myself. Get it together. It's all in your head. I took my coveralls off the hook. Even raising my arms seemed to speed my heart up. This shit had never happened to me. Just yesterday, I was working fine. I managed to get my coveralls on and headed out the berthing, towards the ladderwell (Navy term for steps), headed to the office. I had to stop after getting up the first ladder. I wasn't feeling well at all. Something wasn't right. Slowly, I took 3-4 seconds between steps getting up the second ladderwell. Same feeling. I managed to make it to the office and sit down. Damn, what the hell was causing this? I logged on to the computer. The minute I opened my email, it felt like someone's hand squeezed my heart. I took a deep breath and tried to shake it off. About another 20 seconds went by, and the squeezing feeling occurred again. I just knew shit had gotten real. I rose up slowly from the chair and walked to the gym next door. It was early Sunday morning, so no one was there except for a small group doing PT in the corner. I made it to the gym office where a young cat was sitting in there watching TV.

"A man. Can you call medical?," I told him.

"You aight?," he asked.

At that point, I wanted to cuss his ass out. No, I am not okay. If I was, do you think I would ask you to fucking call medical? However, my heart was spazzing too much for me to lash out at anyone.

"Nah man. My heart. Call medical."

He got on the phone and it seemed my heart was now in overdrive. I got up and stumbled across the gym.

"HELP! MY HEART!" I yelled to the group doing PT. Then, I dropped. I didn't blackout or lose consciousness. I simply dropped, struggling for air. It felt like cinder blocks were on my chest. This was it. The one fear that I always had was coming true. I was gonna die on board a Navy vessel. It's a running joke with sailors that we are willing to die anywhere. Just not on a Navy ship. I'd rather jump overboard and get eaten by a giant squid than die on board a warship. People were around me as I kept gasping for air. I heard over the 1MC:

> "MEDICAL EMERGENCY! MEDICAL EMERGENCY! MEDICAL EMERGENCY IN THE GYM!"

My eyes were open, but wasn't nobody home. The medical team showed up and literally drug me to medical. I don't know if it was will power or stupidity, but my arms were around two people and I literally scooted my feet across the deck plates. As they got me to the medical ward and I lay on that table, my heart still going as fast as a Lamborghini, I thought about a lot of things. What if this was it? Navy doctors weren't what you call exactly trustworthy. Ask anyone in this force. Hell, the whole military for that matter. If you were hurt, 90% of the time, medical had a guaranteed recipe for you. It was called water, motrin and rest. Shit, half the time, if you went up there, many of them acted like it was a damn problem that you came to see them. Now, this doesn't go for all corpsmen (Navy name for hospital workers), especially ones on the green side (Green Side=Marines). When you are literally a Marine's lifeline, you take your job a little bit more seriously than most. It was the ones on the ship that pretended to not give a fuck about people. A few select held up the code when it came to taking care of folks, (like my guy KB) but many didn't. You can say that with a lot of people in the Navy, but when you have people's lives literally in your hands, it's a little bit different. I thought about dying, and really being pissed with God that he would take me at 24. I was surrounded by folks, conscious, but scared was an understatement. In a nutshell, they sent me on my way after an hour, saying that I had an anxiety attack. How the fuck did I have an anxiety attack two days out from Hawaii, and a week and some change away from America? I would've preferred they just said *"We don't give a fuck. We're going home and ain't got time to deal with this shit."* Over the next two days, I was going through it. I couldn't walk up steps without chest pains or a rapid heart rate. This shit was for the birds. When we pulled into Hawaii, despite what had just happened, my chain of command wanted me to partake in the PRT (Physical Readiness Test). Mind you, most commands just waive that shit come deployment, cause we gotta lot of other shit to deal with. Naw. They really expected me to do this. I had gotten a run waiver midway through deployment when my knee popped walking up the well deck ramp. For my civilians, the well deck is where we parked the hovercraft that drove Marines and their equipment to the beach. Later on, I would find out that my meniscus had torn once again. I wasn't running at all, but they really wanted me to do push ups and sit ups. That shit wasn't gonna happen, and I avoided the bullshit thanks to one of my OGs who was there with me, and wasn't gonna let it go down. That October 29th night, despite doctor's orders, I left the ship to stay out in town. I was staying at a now defunct hotel called The Waikiki Wave. It was right next to the now defunct nightclub called Zanzibars. I remember specifically that I was in room 404. The whole way leading up to this moment, it was the same shit of chest pains and rapid fire heart beat at times. By now, I just knew I was gonna die. My mental health was all fucked up. So, with all that had occurred from the previous two days, I just said fuck it. If I was gonna die, then it would be on a mattress, and not on a big piece of steel. I made it to one of my folks from the ship room early in the day and started drinking with them. It was an hour of chillin and enjoying what I thought would be my last moments. As I left, they gave me the bottle of Southern Comfort and I went back to my room. I had a homey who was also staying in the joint with me, as this joint had two big ass bedrooms, a full living room and you could walk out the sliding doors to the flat, fourth floor roof level. Like, I could literally walk out to the edge of the building and smell the spam in the air. Yes, Pacific Islanders fucked with spam. Not the store processed shit. The real shit. I would learn how truly good it was when I moved to Guam, but back to the story. It was no biggie at all for me to let the homey stay in the room with me, seeing how big it was. I chilled for a minute before I drug myself to the elevator, out the hotel and down to the ABC Store. Anyone who has been to Hawaii knows that they got more ABC Stores than people. You found what you needed at the ABC store. If your homie lost his soul in Waikiki, chances were that it would be at one of the ABC Stores. Vanilla Coke was still poppin around this time, so I copped some of that. As I walked through this party vibe of a strip, I can't remember what store I went into. Shit, I don't even remember a full blown liquor store in Waikiki. I found some shit that I knew would get me fucked up. I didn't even know that Hypnotiq was still sold. Had I had some Henny, then you already know what was gonna take place. However, Southern Comfort would do. I walked the streets until I was back in front of my hotel. I looked up and next to me, and saw half the ship going into the club. The other half were standing outside. I kept moving but not before really looking at 'em. I knew that this was it. My mama was gonna get a call in about a day or two that they found her son dead in a hotel room. I made it up to the room and just sat in there, pondering on everything. I washed out one of those two glasses every hotel gave you when have a kitchenette. I pulled some ice out of the freezer. It was something small that made me happy. The freezer here automatically made ice. A cold drank for a cold heart that wasn't functioning correctly. I needed this alone time, so I'm glad fam bam was out and doing whatever. I had gone through some health issues before, but never like this. I poured up the Vanilla Coke and thought about my brother Reggie, from Nashville. This was his go to when it came to mixing drinks. I poured a small hint of the Southern Comfort, and finished it off with a good amount of Hypnotiq. It was late, probably two hours until midnight. I went to my bedroom, with the Japanese doors and shut them. With my heart still racing a mile a minute, and chest pains continuing, I just sipped slowly as I watched TV. Rather, it was watching me. My goal was to sip and drink until I could literally feel nothing. Probably 20 minutes in, I made a call to a shawty from the ship. We clicked on deployment. I was in no condition to walk, let alone have sex. However, when you believe that you are gonna die, you might as well go out with a bang. She didn't answer, so I continued on with my night. Glass full, glass empty, I continued on, well past midnight. I was buzzed beyond belief. I knew the alcohol wasn't good for my heart, but fuck it. I wanted it to stop. That was the whole point. As midnight passed and I was on my millionth glass it seemed, a special came on, highlighting the upcoming presidential election. Who would win? Barack Obama or John McCain? You already know who I was pulling for. I had never seen anyone with my skin color even get that far to be in position to run the country. I was hoping he won, cause I knew that I wasn't gonna be there to see him when he got elected. I stayed up until about two in the morning. I cut the TV off, locked the Japanese doors and hit the lights. This was it. I had enough alcohol to supply the entire United States Navy in my system, my heart was out of whack and I just knew the next time my eyes opened,

I was gonna be looking at Jesus. He was either going to say *"C'mon in fam,"* or *"Take the escalator down."* Either way, I just wanted this earthly life to be over with. I didn't even think about what my partna would have to endure when he came back in and found me dead. Hours later, sometime after six in the morning, I woke up. My chest was on fire, but I was up. Why didn't I die last night? That was the whole purpose. I never told anyone this story in depth until now. My mind was fucked up, as was my spirit. I rose up and sat on the side of the bed, breathing deep, wanting to cry, but I couldn't. God must've really had a purpose for me, cause I knew this shit was gonna be an uphill battle, and I was still at the bottom of the hill. I looked at my empty glass and near empty bottles as I cut the lamp on. I really just tried to commit suicide and it didn't work. Whatever I was gonna do in life, it had to be extraordinary, since The Man Upstairs didn't take me last night. I got out of the bed and opened my Japanese doors. I heard the shower running in the bathroom, so I knew that was the homie getting cleaned up. The early morning light coming through the living room window was glaring through the window and fucking with my eyes at the same time. As I walked into the living room, planning to open the sliding door and walk out on the roof for the crisp, morning Hawaii air, I saw her on the couch.

"Hi," she said.

Immediately, I sobered up. What in the good fuck was this naked White girl doing on my couch? I stood in shock as I just knew I was dreaming. This was the set up for a bad circumstance. I was Black and White people already don't like us as is. Now, White girl, with a nice ass body may I add, was ass naked, on my hotel room couch, chillin. This shit smelled like a rape accusation to me.

"Ummm hi," I said, exactly like that.

I mean she was hella comfortable too. At that moment, the bathroom door opened up and my homeboy walked out with a towel around his waist. He looked up and saw me. I gave him a smirk like bruh, what the fuck was going on last night? He looked back with a grin that said hey, it was grass in the infield, so I played ball. Fast forward to half an hour later. She was gone and I just had one question. I asked him what time he brought her up there. He then replied that it was two of them, but since they saw me sleep, he didn't wanna wake me up, so he had fun. Now, ain't this bout a bitch. I let him stay in the room for free, no money on it and he just had himself a good ole time, while I was out. Maybe, it was the right call. Maybe, a nut woulda killed me. I don't know. All I knew at that point was that brothers from Milwaukee were some wild characters. My mind was still shot, but for a split second, it was stabilized. Fam bam made me realize, there were no friends, like military friends.

As you just read, this story was about a failed suicide attempt. It is the first time I have delved in depth about this experience. As a now 38 year old man who has gone through additional traumas in life, I am here to tell you that it is OK, to not be OK. When you know you are not OK, it is the time that you recognize that help is necessary. I began my therapy journey in 2014, with pre-marriage counseling. At first, the shit was taboo. My biggest question is why am I paying someone else to listen to my problems? However, the more I went through it, the bigger my growth process became. Soon, therapy was damn near as regular as me drinking water. Until this day, I continue to see a therapist. Half of the time, it's not even about shit being wrong. A lot of the time, you are just fine tuning yourself. Suicide, attempting it, or even contemplating it, is not the way to go. Looking back at 2008, I don't regret the action one bit. One, had I not tried it, I wouldn't be here giving you all this story. Two, there is no sense of regret. At that point in time, that was my mindset. I can't change it and wouldn't even if I could because it led to the person I am today. One thing I have learned about choices is to not look back and say what the fuck was I doing? You were doing exactly what you were supposed to be doing because you were either young, you didn't learn to think things through yet, or the situation that arose called for a certain action at that time. I used to party until 4:00 AM. I don't do that anymore. I used to feel a certain way around LGBTQ people. Those days are long gone since I met my friend Ericka back in 2009. I used to eat wings every Sunday (Like no shit every Sunday). I damn sure don't do that anymore because I am more health conscious. I even hated getting pampered because I thought many years ago that men don't do that. Now, you can catch me sippin a cranberry juice as someone works on these raggedy ass feet of mine. You can catch me at the sports medicine clinic, getting a deep tissue massage and my back cracked like a bag of pork skins. Point being is this. Don't be ashamed of how you were. It was how you were wired at the time. There is no need for apologies (This does not count for deadbeat mothers and fathers. Fuck you always). The biggest thing I want to tell you is that you have purpose. You are here for a reason. When you think you don't have a purpose, take a deep breath. You see how you are still breathing? That means you aren't done living your life. Below, before we get into the next story of mental health challenges, I have listed some (Not all) resources and phone numbers that offer mental health assistance.

24/7 Crisis Hotline: 988 Suicide & Crisis Lifeline
`https://988lifeline.org/`
If you or someone you know is struggling or in crisis, help is available. Call or text 988 or chat `988lifeline.org`. Veterans, press 1 when calling.

Crisis Text Line
`http://www.crisistextline.org/`
Text TALK to 741-741 to text with a trained crisis counselor from the Crisis Text Line for free, 24/7

Veterans Crisis Line
`https://www.veteranscrisisline.net/`
Send a text to 838255

Vets4Warriors
`http://vets4warriors.com/`

National Suicide Prevention Lifeline:
1-800-273-8255 to speak to someone if you or a loved one is considering harming themselves or experiencing emotional distress.

Anxiety and Depression Association of America:
240-485-1001

Mental Health America: 703 684-7722

— 5 —

ONE SHOT AWAY

Ahh the joys of being young. Man, it is so crazy how much growth we do over time. I sit back right now thinking about all the stupid shit I spent money on when I was between the ages of 18-22. For starters, let's start right after boot camp. True, my hometown was nothing but a 45 minute to an hour drive for Naval Training Center Great Lakes. I could've gone home and got clothes for days (I eventually did). Yet when you've been wearing a raggedy ass uniform for nine weeks straight, the first piece of normalcy that is available, you are gonna take advantage of it. In the case of young, dumb, uneducated sailors, it was fly ass urban gear. When you went into the Navy Exchange on base, which is the name for their convenience store, the first thing you see entering through the sliding doors was all the urban gear. Now, this was January of 2003, so Platinum Fubu, Phat Farm, Akademiks and Enyce were still hanging on for dear life. Champion was still in Walmart and a target for a severe roasting. PUMA wasn't on the grand scale like they are now, outside of the soccer scene in Europe and certain areas in America. Man, it's crazy how times change. When I got out of boot camp, I checked my bank balance. I had $1,400. You couldn't tell me shit. At 18 years old, you might as well have called me a millionaire. I quickly learned, however, when I left school and got to California, that I didn't have any real money. As an E-3, I made $400 per paycheck. Now, consider that I had a car note and a phone, and you're looking at almost $300 down the drain. If it wasn't for that enlistment bonus, a brother would've been hurting. Now, the money was one thing. Truthfully, it still didn't keep me from spending the shit. I was 19 when I touched down in San Diego, back in November of 2003. 19 year olds are gonna do what 19 year olds are gonna do. On payday Fridays, I remember a gang of us used to wait by the ATM on base, by the ships, waiting for nine o'clock to hit. That was midnight on the East Coast, and that's when the money hit our accounts. I mean the line was long. A bunch of 18-21 year olds with nothing else to do but spend the shit. Most of us were simply pulling out some funds to hit the club. Back in those days, you had a good selection for the 18 and up joints in San Diego. As an adult, ain't no way in hell I would be caught in any of those (I say me cause it's some old heads who like to creep around those joints). However, the big staple for all of us in the city was Werk Dat, Twerk Dat. Someone who was in San Diego at that time just read this and laughed out loud, I know. This joint was in a spot called Antonio's, on 7th and C downtown. When you walked upstairs and inside the actual club, you saw that it had two sides. One was for 18 and up. The other was for 21 and up. Still, both sides were about the size of a small apartment. Every Saturday, at least 1,000 of us packed in there to dance, pull some women and fight. Those were the three guaranteed things that happened there. The three guaranteed groups in there were Bloods, Crips and military. You can just imagine how it was. No matter what though, we stayed there every Saturday regardless. I was a party head, and still am to this day. As with everything, the older you get, the more you adjust and change your scenery. Now, it's the VFW where I have a drink and listen to old, retired vets talk about how much cooch they slayed in their heyday, while eating some red snapper. More important than anything else, no matter where I go, I have and always have been someone who wanted to control my movements. I never liked depending on someone else to get somewhere. Now, don't mistake this as being ashamed to ask for help. That, I am not. What I mean by controlling my movements is this. If I'm going somewhere, I wanna be in my car. When I am ready to leave, then I can take my Black ass home and chill. I'm pretty consistent with all that I do in life, but the one time I deviated from the norm, it put my life on the line and fucked with me for the rest of my life. Welcome to the mental health battle with sheer stupidity. It's funny. This is the shit that people don't talk about. What happens when the uniform comes off and you return to being you. It was August of 2006. I had just transferred up to Camp Pendleton, which was 30 minutes north of San Diego. Over the years, I grew to love the Oceanside area. However, at that time, not so much. You would've thought that San Diego was 5 hours from O'side instead of 30 minutes. I wanted to be back down in the city where everything was cracking.

Every weekend, I would drive my raggedy ass Mercury Sable back down to the city and stay with a friend. I would rip and run, party, chase women and not have a care in the world, especially seeing how gas did not cost you a pint of blood and two of your fingers. This weekend, I was staying with a homie and his roommate. In short, there was a house party in the city tonight. Now, this shit was down in an area called Eastlake. It was off the 805 freeway, in a chill part of the city. Nice homes, low crime and no worries. As always, Friday was an international half day in the military, so I caught a good 2-3 hour nap when I got to the homie house. I only had one thing on my mind. Females. Every time I went to the club or a function, my goal was at least to pull one, get a number, something. I hated going home empty handed. Night time came and we rolled out. My guy was like me. He liked to be on time. As y'all know, don't nobody show up to a Black folks function when it first starts. However, we had that Play from House Party mindset. If you don't know what I'm referring to, let me explain. (Y'all who have seen this cult classic a million times know where I'm going with this). Bilal was Play's homey, and the DJ for the party. Hours prior to the start of it, Play had swung by and loaded Bilal's DJ stuff into his car. He told him that he would drop it off and come back. Long story short, Play's ass didn't come back for hours, so Bilal was pissed off. When he did finally show back up, he had a car full of females and Bilal said he wasn't going, Then, Play told him the smartest shit you could ever tell a teenager.

"They keep asking. Who's the DJ? So I brought them over here so you could get a jumpstart before the other brothers."

That was the prime reason we were early. We wanted to get on before all the competition entered. As we entered the house that night, it was a typical experience. Few bruhs, but a good amount of females. Fam bam knew everyone and I didn't, so I just kind of rolled with things and let him introduce me. I chopped it up with a few shorties, and got better acquainted with them. The house filled up after some time, and I found myself in the kitchen. The host asked everyone to come into the living room. I aint gone say what was going down in the living room, but I saw a shorty sitting outside on the patio, through the sliding door in the kitchen. I went out and told her that her people said for her to come inside. She was clearly not in a good mood, and in a nice way, she told me to fuck off. It was totally fine by me. As I slid the door shut, she yelled for me to stop and asked if I could come outside. Quick judgment saw that she was fine as ever, so I went. That's the honest truth. I didn't give a shit about her crocodile tears. She was fine and thick like a pot of grits with salt, pepper and butter. We began talking and I truly didn't care what she was going through. She was fine and I was just hoping to say the right things so I could possibly slide into something at a later day and time. If you're mad at what I just said, so be it. I was 21, and thinking with your dick before your brain is something that most 21 year old men did. As I stated at the end of the suicide story. You think or act how you are supposed to act and think at certain points in life. We ended up laughing and joking after a while, and walked back in the house hand in hand. As we came back into the house, a guy was staring at me and her. With some hesitation, he then walked up and asked if he could talk to her. In typical fashion, I let her hand go so she could do her thing. She wasn't mine and I wasn't cuffin. Plus, I got the number already, so I was good. The next 30 minutes were entertainment, and then, shit got real. I don't know what had happened upstairs (I do, but as stated earlier, everything ain't meant to be told), but two people came down and went outside arguing. A man and a woman. At this point, I found myself in what you would call the yard area of the house. I say that because if you live in California, you know having a yard is rare, seeing how the homes in the city are damn near built on top of each other. I was back there with some of the guys and the dude up front had pulled out a pistol. Fucking great, I thought. Here I am, a week before my 22nd birthday, and I was finna be part of a movie that I didn't wanna be a star in. Two deployments in a hostile area, and I came back without a scratch on me. Now, I was on the verge of getting blasted in San Diego. Here was the mental battle of "I know you fucking lying." This shit wasn't supposed to happen. Not in this part of the city. Chatter was going on as we looked over the gate and these folks were still arguing in the street, with more folks now outside. A few minutes later, homey took off with ol girl in the truck. I ain't know what the fuck had just happened. All I know is that I was ready to bounce. I didn't need this shit. I was still at the point where I thought the smallest incident would get me kicked out the Navy and I would have nothing to fall back on. I had started school at this point, but was nowhere near finished. My money was still funny and my credit wasn't gonna get it. Truth be told, I needed this shit in my life right now. As we made our way to the front of the house, one of the folks was talking to them inside the vehicle, on speaker phone. All you heard was commotion, arguing and then a sound that I will never forget. Metal was crunching. It was horrifying to hear. My man ran to his car and I followed. I wasn't about to let my guy go anywhere by himself. It wasn't even about him being my ride at this point. I was in the mode that if my homie was involved, I am too. Everyone else followed suit and we got to riding around this part of the city, looking for 'em, not knowing what we were gonna run into. After a few minutes, someone called my man's phone, and he took off on East Orange Avenue. When we pulled up, the sight I saw fucked me up. One of the girls was on the ground, in the middle of the road, bleeding badly. Two dudes were sitting off to the side of the road like shit hadn't happened. The truck was upside down. Immediately, amid the commotion, I took off my shirt and threw it over shorty that was laying in the middle of the street. I was scared to apply pressure anywhere because I didn't know how bad she had been injured. All I knew was that I had to do something to at least help her survive. When you are 21 years old and haven't navigated life, you truly dont know what the fuck to do in a dire situation. You learn by trial and error. A huge fight then erupted in the street between the dudes from the truck and some of the women that were at the party. They tried to beat the breaks off those boys. Police and EMTs showed up and I just knew like a muthafucka that we were all going to jail. I didn't need this. None of us did. Luckily, by the grace of God, 12 got control of folks and they were able to get babygirl girl in the ambulance, and rushed her to the hospital. As we were now at Scripps Hospital in an undisclosed part of the city, with tensions still on high, I called my grandma. When things got hard in life, that's who I called. It was past midnight here and two in the morning back in Indiana. I must've called three times before she answered. Immediately, I started crying and pleading to her that I was in trouble. Hell, I probably had her thinking that I was in an accident for the first few minutes. Finally, she calmed me down and I was able to tell her the story. She broke me down gently and in a way that only a grandmother could. I asked myself one question at that time. How would my grandmother react if she saw me in a casket? I just knew this shit wasn't over. With all the drama that occured at that house, this shit was just getting started. Yea, I am intentionally leaving out details in this story, along with names. All that doesn't need to be discussed. The next day, rather later the next afternoon, we were in a car four deep, in front of an apartment complex. Homeboy inside had guns supposedly and I knew someone in the whip I was in had 'em. We all try to pretend like we are the toughest beings walking the earth at that age. However, tough ain't putting yourself in situations to show you gangster. Ever since that August 8, 2006 day, I have dreaded when that day shows up on the calendar. It's a memory that still sticks with me to this day. I saw too much that day. A bloody body in the middle of the road. A pissed off father at a hospital. I didn't need much more. See, everyone tends to forget that we are human as well. That uniform is a cloud that blocks the sunshine for a few hours of the day. Every now and then, the sun is involved in an eclipse and full darkness comes over us. This was my darkness. Never again have I ever driven with someone else, anywhere else, unless it's been my brother D, and that's cause I know for a fact that he would never put me in a predicament that would risk my life. I love life, but I remember the first true time as an adult that I thought it would end. So many different scenarios could have played out that night. Dude could have dumped on us at the house, or on the road when we got to the accident scene. Me and fam could have flipped and crashed our damn selves because he was driving like a bat out of hell, with good reason. There were so many things that could've gone wrong that night. I know could've, should've, and would've doesn't matter. Neither does if. However, those four words can teach us valuable lessons that we can learn from. I learned the most valuable one that night. Enjoy every day as if it is your last. I'm just glad God decided to kick me in the ass that night instead of saying come on home. More so, my grandmother didn't have to bury her grandson.

— 6 —

WHEN THE GRIM REAPER CALLS

Before I get into what will indeed be the longest, most sensitive and angriest part of this book, I wanna write a letter that I never got the chance to write. Please, bare with me as I express my thoughts:

Dear homies that I lost along the way. I am sorry that I couldn't save you. I am sorry that I couldn't be there during the roughest parts of your lives. Man, It's crazy. We do this military shit through four different worlds. In the Navy, we cruise on these steel ships. Funny how a lot of the world thinks that the Navy is all stationed on one big ass ship. I stay telling folks that I don't know y'all family members. If they are in Virginia, I'm in San Diego and we ain't meeting up everyday. You Army guys play in dirt and we've kicked your ass in football way more times than you could ever imagine, but you are still family. Marines, the Navy children, but some of the toughest people I have ever met. Air Force, the spoiled rotten children of the military. I sometimes debate on whether or not y'all are really the military. You drink purified water, live on resort style bases and will get time off for getting a paper cut. However, I still love you. I know these jokes are all too common. You know what else is? People not caring when we go through tough times. Y'all know me. I never try to put myself above someone else. Only those of us who serve, however, will understand this shit. Being mentally fucked up in a war zone. Being told to suck it up at times. Going to someone for help, only to be told to get some rest and you'll be fine in the morning. Then, when the bullet leaves the gun and becomes a permanent part of us, it's always, "What could we have done to save them?" They play that role as if they give two fucks. Especially the higher ups. I know it ain't too many that truly care about us once they hit those O ranks. It's some real ones that trickle into those ranks, but far more fake ones. So, I'm sorry. I take full responsibility. If I could hug all of your parents, I would. I only hugged one person's parents after their death and that is what I am about to tell the world. The experience that killed my soul. I promise you, as y'all look down on me right now, I will spill every detail. I will spill every emotion. People need to know how serious this shit is. I love y'all man. Even if I didn't know you personally. Sincerely, your veteran brother, Joe.

DISCLAIMER: AS WITH EVERY PART OF THIS BOOK, WHAT YOU ARE ABOUT TO READ ARE TRUE EVENTS. I HAVE PURPOSELY LEFT OUT OR CHANGED NAMES. ONE, OUT OF SHEER RESPECT FOR SOME PEOPLE. TWO, BECAUSE MENTIONING SOME PEOPLE'S NAME'S WILL TRIGGER ME. THIS IS A STORY ABOUT MEETING DEATH FACE TO FACE.

The day was May 26, 2016. I was three months deep into deployment on board the USS Harpers Ferry. Trust me when I tell you. Fourteen years in the Navy, and this was by far the worst fucking command that I had ever encountered. I would have rather been Tom Hanks stuck on that island than be here. Now, let's get into how this place became the worst place that I ever encountered in my life. I have to take you back to understand the journey leading up to this. As you know from reading earlier, I started off on the USS NIMITZ. After three years in a good space, where people looked out for each other, I transferred to Assault Craft Unit 5. In both these places, I had some great leaders from top to bottom. I mean, when I say they looked out for their people, they truly did. When I left ACU-5 in 2009, I transferred to 32nd Street Naval Station to a command called FISC (Fleet Industrial Support Center) San Diego, which would later be renamed FLCSD (Fleet Logistics Center San Diego). Here, I became a certified hazmat technician and journeyman instructor, meaning I could teach pretty much anything. This command, which was majority civilians, still embraced everyone. We were a team. My boss on the civilian side, became more like a big brother. My Division Officer, I can never forget him. He was my boss on the military side. Look out for his people? That shit was disrespectful to say. Let me put it to you like this. If he were a surgeon, he would fix the problem that he knew of and then get other bad shit out of you just because. He was a great damn leader. No, excuse me. He was **FUCKING PHENOMENAL!** After three years at that place, I took my talents to Guam for two years on board the USS Frank Cable. I wasn't supposed to go there. I was originally slated to head to Diego Garcia. However, with one of my homies in charge of my detailing, he cut a drug deal, got me a shit ton of extra money and I was Gucci. To that partner, I am forever indebted to you. Now, back to Frank Cable. My Senior Chief. Mr. No matter what happens, I'm happily married and I'm going to get laid when I get home. Then, Big Chief Don. The Last Don. **WHAT!!!** Good leaders? Naw. They were great fucking leaders!!! Add in the Triple OG of the Boat, Senior T, who is hands down the most thugged out muthafucka I have ever met in the Navy, and you had the damn 1996 Bulls for a khaki brigade. The day I went in for my Officer of the Deck board, I remember Senior T looked at my PQS and asked one question.

"Have you stood OOD before?"

"Three years Senior."

"So what the fuck you doing here for a board? You know what the fuck you doing. Just know if you fuck up anytime up there, I'm on yo ass. Got it?"

"We good."

"Aight. We 219 baby. Dont fuck this up."

219 was the area code for Northwest Indiana where I'm from. I'm from East Chicago, a city of about 30,000. He was from Gary, which at its peak had 200,000 something residents. It was a fraction of itself now, but still, thugged out nonetheless. We understood each other and never had any issues. My two years there were great. Outside of the random weekend nights where a few people did stupid shit and I was on watch (Brawls at the barracks, brawls at the club out in town), this was something that Hollywood couldn't

script. Four different commands with four different work environments that were not toxic at all. How the fuck did I get so lucky? I didn't even think about the statement that all good things come to an end, cause I would've jinxed myself. After leaving Guam in early 2014, I went to work at Naval Hospital Balboa in San Diego for 6 months. I had to get all my tests done on my heart to make sure that I was still good to go. I had no major medical issues in Guam, but a few doctors thought differently. It almost got me separated, but two dope ass Captains and one gangster ass Master Chief gave the Navy the middle finger and told them, this our guy. We gone ride for him, vouch for him and we want all the smoke. Back to the story. One thing I will give the cardiologists that I have encountered during my time, especially at Balboa, they always took care of your boy. I wish I could've stayed there forever. I didn't know that June 25, 2014 would be the beginning of a four year trek through hell with the devil sitting courtside enjoying the misery. In my shop, I had no issues with who I worked with. My guy Big Country, he was solid. We would end up forging a relationship that goes beyond the Navy. My guys were good. A few of those other ones though, not so much. You know when you meet folks for the first time, they always say that you are meeting their representative and not the true them. After those four years, and they know who they are, I was glad to say fuck em and know that if they ever say anything to me on the other side, its gone be fuck em again. Let's jump to 2015. It was that time where we were gonna be starting work ups for our upcoming deployment at the beginning of 2016. Lemme keep this shit a buck 50. The

Engineering boss, he truly didn't give two fucks about us. All this man knew was work, work, work, and more work, without any regards to anyone's personal well being. Work days were long, and when you combine them with all the days we spent out to sea, on top of duty, we were drained as an Engineering Department. I mean totally drained. I was married at the time. When I got home, which most nights was after 6:00 PM, I had the energy to eat, shower and sleep. I played with my dog for a lil bit, but I know I'd have to be back up at 4:30 AM the next morning, just so I could get to base on time, park and start my day at 5:45 AM. Trust me, anyone stationed on any Navy base will tell you that parking is pure ass. I'm exaggerating this a little bit, but if the Navy had 5,000 people in one work space, they'd create a parking lot with 500 stalls and tell you to figure it out. For most of the year, we were on three section duty, because the ship had to be "deployment ready." Oh, let me not mention the one, two, three, maybe four times that we came home early because machinery fucked up as soon as we left out to sea, and we had to pull right back into port. Get new shit? Nah. It was the mindset of putting more stress on your engineers for some shit that was on its last legs. By the end of that year, we had multiple suicide attempts, with one guy being lucky that his wife walked in the garage when she did. He had supposedly let his car run with the garage door closed and just laid in there. Pregnancies? Man I knew it was bad when one of the baddest women on the planet got pregnant. Not on the ship. On the planet. I won't say her name, but my God she was too damn pretty to be in the service. She should've been in someone's magazine, making six figures for modeling. She was Haitian, brown skinned and as the old heads used to say, I'd drink her bath water. She wasn't the only one who got pregnant and got out of that hell hole. A few more of 'em did as well and I couldn't blame them. There was no empathy for any of us. From top to bottom, it was work, work and work. Come January 2016, we had what they called INSURV (Inspection and Survey). In lamen terms, no matter what the fuck was wrong with this ship, we were still leaving on February 11th. They passed us, supposedly. Truth be told, we were fucked worse than the damn boat itself. We were overworked, tired and maybe was home for a quarter of the previous year, and that's being nice. People barely saw their kids, their wives/husbands, and had time to handle personal matters. The shit was crazy. And what did we get two weeks before we had to leave? Four fucking days off. Four days. I know that's all they could afford, but did they think it was gonna make us feel any better? The last straw was having us come in the night before we deployed. So wait a fucking minute? You telling me that a whole crew who put in work, gave their all at their lowest points, who barely had time to do anything over the last year, couldn't even get a full fucking night with thier spouses, kids or just to relax? I had never experienced any shit like this on my previous three deployments. The first one, I flew out and met the ship. The second one, we came in at 6:00 AM and we left at about 10:00 AM. The third one, we were ready to bounce at 11:00 AM, but it ended up being a two hour time delay, so we left at 1:00 PM. In between those two hours, we made a last dash to the liquor store and stocked up on every drink we could imagine. Oh yea. That LCAC had its perks. Now, I was on the ghost ship. A bunch of people who had lost their souls because they had nothing left to give, and couldn't even get one final night with the ones they loved the most. We left that Thursday morning with a mental strain of epic proportion. Go to Hawaii for a few days like everyone else on the West Coast? Nah. That shit wasn't happening on this go around. Instead, we floated for 44 days around South Korea and the South China Sea. It was colder than a muthafucka too. Like these muthafuckas couldn't even give us Hawaii. That was the first stop which basically signaled hey, buy the shit you may have forgotten and turn up before the bullshit months that lie ahead. Naw. Not Harpers Ferry. This rust bucket was hell bent on saying fuck yo life. Eventually, we pulled into Hong Kong, which I had been to before. It was a good wind down, but we were still frustrated. Now, when we pulled into Seychelles, East Africa, in April, things changed for those few days. I was in the Motherland. It was more than just touching African soil. It was the mental break that I needed. Hell, the one that we all needed. Truthfully, this boat had taken my heart and didn't give that muthafucka back. I operated at 50% which was something that I never did my entire career. If they weren't gonna give back to me and my guys, why was I gonna give my full effort? People can say what they want and I personally don't give a fuck. I know the majority of people who have run across me in the Navy will tell you that Mac is a hard worker. Mac keeps it a buck 50 at all times. Mac looks out for folks. Here, that shit didn't apply. I kind of just passed the time by. For those four days in Seychelles, I got my dance on, ate well and saw some of the finest African sisters that I had ever seen in my life. That place was paradise. The last night there, they took the whole ship back on a three story mega yacht. When I say we were wilding the fuck out, we were wilding the fuck out. I mean near fights. People throwing up from the third deck onto people on the first deck. Dance battles, ratchetness, all that. We were truly in that sailor element. Now that I think about it, this place probably saved us all. We were so used to the non stop grind, that when we finally got the chance to let our hair down, we dropped the whole fucking wig. Life went back to normal after that, until that fateful day. (Warning: If you dont wanna become sad or angry, just go to the next chapter of this book) It started off normal. I got my ass up at 5:30 AM like I did every morning and made that walk to that oh so ever immaculate bathroom of ours that we called a head. I did my do, brushed my teeth and headed back to the berthing, to put my coveralls on. (I shower at night before bed. So no, a brother was not funky). I headed downstairs to the galley. At this point, I was good with all the cooks. I always started shit with them later in the day, but never in the morning. That's an unwritten rule. Black folks will tell you straight up. It's too early for that shit. I was the same way. I went behind the line before everyone else got down here and made my eggs as usual. What did the cooks get in return? Whatever the hell they wanted. If they shit broke, I made sure it got handled. You didn't fuck with people who made yo food. As I said earlier in the book, I know they catch a lot of flack for quote on quote "Navy meals," but the majority of the time, they were good. I ain't saying our meals were Benihana's, but it damn sure wasn't Taco Bell.

So to Bama, Gizzle, Steve-O and Gunz, thank you. I made my ham and cheese omelet and went into the First Class Mess where I was by myself. As usual, Groundhog Day was on, as one of the officers literally wanted this shit played everyday. I tried to watch sports, but AFN (Armed Forces Network) was going in and out because our satellite signal was fucked up, which was 99% of the time. I chilled in here for a good minute, as I knew some of my guys would be waking up and making their way in here. It was one of our wind downs before we had to start the bullshit for the day. If it was one thing this ship could provide on a constant basis, it was a daily dose of bullshit. 6:45 AM came around and I headed downstairs to my shop. Muster was the first thing on the agenda. I logged onto the computer, going to our muster folder so I could print out the sheet for quarters. The bells on the ship rang over the 1MC. Great, I thought. Gonna start the day off with a fire or a flood. Or, if it's my luck, it'll be a toxic gas casualty, where we gotta put on the breathing apparatus, take the six gas analyzer down and find shit leaking from a pipe in some random space. If only one of those were called.

"MAN DOWN! MAN DOWN! MAN DOWN IN THE ARMORY!"

Wow. There was something that you didn't hear everyday. Now, I had been at a command previously where someone had shot himself in the foot trying to clear his weapon after coming off watch. Here's the thing, however. We were out on deployment. Cats didn't walk around with the 9MM. Not even MAC (Master at Arms Chief). Someone had to have been in there playing with the joint, probably trying to take pics for the gram and it went off. It wasn't surprising though, especially in this day and age. Big Country had just happened to walk in as I was getting up from the chair to report down there. I was a fire marshal, and us, along with the medical team, had to respond to shit like that. He told me he got it and to continue with the muster. Cool, I thought. I went on about my business, definitely getting ready to hear a story at lunch time. Then, I heard eight words that would change my life forever.

"SET RIVER CITY ONE! SET RIVER CITY ONE!"

At that moment, I knew it was serious. River city is when they shut down all communications going off the ship. Someone was dead. The question was, who? Immediately, I started to think about all my GM (Gunners Mate) partners. I ran 'em through my head. They had access to the armory, but ain't no way it would've been them. FC's had access, but the two I knew, nah. My quick guess was someone pinged someone for the code, someone gave it to them, not thinking they would do anything, and they blasted themselves. I sat in my shop as some of my other guys came in. They were asking what's going on. I told 'em I didn't know, but I knew it was serious. We all kind of sat around in the shop. After a while, I needed to get the scoop. I knew not to go down to the armory, so I headed upstairs. As I hit the 03 level, I saw one of the homies in the P-Way, balling tears.

"WHO FAM! TELL ME WHO!"

Now, I was scared as fuck. Not trying to sound desensitized, but had I seen someone I really didn't rock with, I may have just passed 'em and not said anything. Or, maybe just hit them with a "You good?" Naw, this was one of my dudes. Someone who was like a brother. I asked him again who? The word Cam came out of his mouth. I backed up. I knew he was lying. I just knew this muthafucka was lying. This shit wasn't happening. Not my lil brother Cam. Nah. Anyone but Cam (not wishing death on anyone, but this was said due to the nature of our relationship). I know damn well it wasn't him. I walked away. Distraught, but telling myself they got the wrong person. Not the 19 year old kid who came to our men's group meetings in the medical ward once or twice a week, just to discuss manhood and growth. Not the 19 year old who I had just prayed with on the flight deck the day before. Fuck no. I knew this shit was a lie. I walked around the ship a little bit before returning to the shop. Country wasn't back, but the entire division was now in there. Most of them, I got along with. Some, I didn't. However, right now, we were all fucked up and quiet because we didn't know who it was. Over the 1MC, they called all the Chiefs to the mess. That's the E7 through E9 ranks of the Navy for those that don't know. I already knew what was coming. The old man (Commanding Officer) probably was gonna tell them who it was, and have them break the news to us. This wasn't my first rodeo. It was, however, my first one as a leader. An E-6 with almost 15 years in. My Chief came into the shop, along with the other khakis probably 20-30 minutes after they called them up. We all sat there in nervous anticipation. I remember exactly where I was sitting.

On top of the metal shear, feet kicked up on something.

"SN ******* HAS DIED."**

That's all I needed to hear as I immediately let out a howl and ran into our other shop. I hadn't cried this hard since my grandmother had died. At least she got to live a full existence, have a bunch of kids, grandchildren and leave a lasting legacy on this earth. This kid hadn't even started life. The mood was somber. Not just for me, but for this entire ship. As I left the shop eventually, the day sort of went into a standstill. One by one, I would run into people crying and I did my best to console them. Everyone has experienced loss and I get that. For us, we lost one of ours in the same place that we live. We couldn't go anywhere and de-stress. The same walls that saw him die were the same walls that saw us sleep. It wasn't like we could just say hey, I'm gonna go take a swim to shore and I'll be back at 8:30 PM tonight. This shit was nerve wracking. However, the worst part about it all was that there wasn't a mass work stoppage. It was depending on the department. Now, I understand that watch stations had to be manned, so there were some jobs that had to keep going, like in the engineering spaces. I understand that. However, the shit should have been emergencies only. The crew needed at least the rest of the day to process this. Nah. That shit wasn't happening and what happened next let me know that I could never trust military leadership at a high level ever again. I probably say about one o'clock, the CO gathered us all in the Vehicle Turning Area. To sum it up, it was a wide open space where we usually had Marine vehicles parked. The crew was down here, somber and sad. As he began to talk, it went into the same ole tragedy story. This was the shit that pissed me off. It seemed like a lot of higher ranking people I met in my time didn't know how to speak. It was certain shit that you said and didn't say. What took the cake was when in a nutshell, he basically said the show must go on. The fuck do you mean? Now, I'm paraphrasing, but that's what was said in a nutshell. It was a bunch of fluff, and to tell a depressed crew we have to keep going, that was all the way fucked up. No shit. We all knew that life had to continue, but this wasn't something that you say at this time. One of my partners was beyond upset. He was known as the Triple OG of the boat. If some shit did happen, you'd have to see him whether it was for just one on one talk, or an all out ass chewing. God forbid you see his hands. He was not to be fucked with. He had the respect of **EVERYBODY.** So, when he got upset, I already knew the crew was taking on his energy, along with the energy that they already had themselves. From that day on, I didn't care about the ship, or the Navy as a whole. I had truly started to question whether all the time that I had invested was worth it. All this off of one incident. I literally felt like I was in hell for these last two years. Now, the devil was just sitting on the flight deck laughing at my ass. Depression started to set in as I started to feel solely responsible for the death of this young man. The worst was yet to come. I had to do the ballast operation for his body to be carried off the boat. To explain to my civilian counterparts, a ballast op is when we fill one part of the ship with water, so we can kick out a mini craft called a LCU (Landing Craft Unit) or an LCAC (Landing Craft Air Cushion). Picture the war version of some yachts.

The LCU was gonna be the vehicle that took my guy to shore one last time. In truth, I didn't have to do the ballast op, as my DCA (Damage Control Officer), the only guy I really trusted, gave me the option to chill. I couldn't let it ride. This was my little brother. If there was one op that I was gonna make sure that the well deck was right, it was gonna be this one. I can't remember the amount of water needed to kick an LCU out, but I think it's eight feet at the sill. Fellow navy personnel forgive me, but I'm retired and it's been damn near five years since I've been in ballast control. The well deck was set and the boss man gave me permission to head down and see him off. I joined the rest of the distraught crew as we formed two lines parallel to each other, facing each other. As the bells rung, and the IMC call was made, SN *********, DEPARTING, the tears started to flow. Chaps led the procession of those who were carrying him on a stretcher, with a flag covering his body. The closer it got to me, the more the tears started to come down. I was raised in the funeral home (my dad was a mortician), but nothing in my life prepared me for this. As his body crossed directly in front of me, I lost my shit. I bent over and completely let it all out. The only refuge I had was my RP1 homeboy who consoled me. This was the shit that made us military brethren have the strong bonds that we had. We may not talk to each other, we may do. But, when shit hits the fan, and its just us out there, you'd be amazed at what the fuck happens. I've seen cold hearted racists hug people of a different color, or fight beside them in a casualty. I've seen straight folks combat a flood next to gays, knowing that they didn't agree with the lifestyle, nor get along with the person.

This was what made our organization what it was. Would people go back to their ways after something dramatic or traumatic happened? More than likely so. When everyone's life is on the line, however, all that shit goes out the window. We are trying to get home, whether it's to spouses, children or just to the comforts of a bed that you don't have to worry about falling out of. Months would pass and I would start to see that the career that I once enjoyed wasn't enjoyable anymore. I really contemplated whether or not I wanted to finish out. Fourteen years deep in 2016, and the last six were really looking like I was gonna say fuck it. About two days after Cam died, my name was called over the 1MC. I was called to go to the administrative office. This was strange. I had partners up there that I fucked with heavy, including the Triple OG of the boat. If I was needed, they usually called the shop themselves. I walked upstairs to the 04 level where their office was. I knocked and entered, because something was telling me that this was gone be some shit. As soon as I entered, my partner from the East Coast, who always kept it a buck 50 with me, had a look in his eyes that didn't sit right with me. It was an uncomfortable look. A bald headed white dude was next to him, along with another gentleman.

"HT1. I'm LN1."

LN, aka Legalmen. If you were seeing the legal man in the Navy, that meant one of three things. One, you either did something bad and fucked up. I knew I hadn't done anything, so that was off the list. Two, you were trying to probably learn your rights and options about a divorce, charges or something else that pertained to your personal life. I didn't have to worry about that either. Three, you were somehow connected to an incident that happened. I knew the first two did not apply to me, so I already knew this had something to do with Cam. He asked me if we could go somewhere and talk. I walked him down to Damage Control Central, or as they called it there, CCS (Command Control Station). I told the watchstander, one of my juniors, to leave for about 30 minutes. As I figured, it was centered around the homey who died. I'm not gonna tell you how my name was connected to him, but just know I am forever haunted by what I heard. We talked for 30 minutes about my connections to him, how I observed him on a daily basis, all that. It was a hard conversation, but a necessary one, where I realized where a lot of the mental trauma of the young folks came in. These so-called leaders, the E6 (my peer group) and above, truly didn't give a shit about the sailors. I wasn't in that group, as I didn't have a reputation of fucking people over, but to know that these were the people I was serving with, it upset me. I couldn't go back to the old Joe. Old Joe was a hot head. I've lost fights before. That's nothing. I've won them too. My early Navy days, all that I'm gone talk to you any kind of way because we under Uncle Sam's protection plan shit wasn't gonna fly. To deviate from the story real quick, in 2008, I drove 50 miles from Camp Pendleton to a base on North Island to confront a sailor about some money he owed and said he wasn't going to pay me. In 2010, a retired Master Chief, who thought she was still active duty, tried to run it up on me, on base. As I told her that day. You want my CMC (Command Master Chief) number, I'll give it to you. Call him. I don't give a shit. You won't talk to me any old way, especially now that you are a civilian. That anchor with stars doesn't mean shit. I don't care how many of the San Diego Chargers she had fucked while she was out here (It was a lot). Talk to them niggas like that. Not me. Lets not even start on 2004 when I blasted a dude bout his wife getting fucked by Jody while we were on the ship in front of everybody. Yea, I was pretty nutty in my day. Thank God for growth. Back to the story at hand. I couldn't look at anyone the same, except for the few in my circle that I knew were solid. Anyone else, it was difficult to obtain their trust. You can say I wasn't the best sailor. You can say I wasn't the best at my job. What nobody can ever say is that I wasn't a real one. That's how leadership is effective. Don't try to be a good leader. Be a real one. When you try to be good, you try to cater to everyone's needs. When you are a real leader, you just keep it a buck with everyone, regardless if it's good or bad. I left DC Central that day with my mental health on -5, and it carried for the rest of deployment. I saw and experienced shit no person should ever have to experience, along with 900 other Sailors and Marines. To say 2014 to 2018 were the most difficult years of my life, that shit was an understatement. The icing on the cake was when I met his parents, along with a partner of mine, and the Captain said he went down to see his body. When I found out from my guy, who was the responding medic, that he told this man's parents a bold face lie, that was it. I washed my hands with the ship, the Navy and everything else military related. The only way my mental health was gonna come back to a strong point, was getting off this damn boat. Crazy enough, the day I left was the same day that I was flying out for my father's wake, as he died while I was out to sea the week prior in May 2018. From start to finish, I literally resided in hell for the last four years. Joe was non-existent, cause I didn't know the man that I saw in the mirror anymore. So, just know when you see a veteran, know that you will never know everything, because everything ain't meant to be told. Some things we are gonna take to the grave, because no one should have to bear the burden of what destroyed us mentally. 22 a day. It's a reason for that, and the leadership on Cell Block 49 were in a league of their own. To everyone who suffered with me, just know, I'm always here. Bama, thanks for the smoke in Jordan when I needed to decompress. Z, thanks for not beating my ass when my anger for other shit took over my words. All you were doing was asking a question and I had no reason to come sideways. Even when I apologized to you, in your vicinity, I deserved to get my ass beat had you swung. Lo, you were smart to leave the Navy. Your husband, keep him. To know we spent less than 100 days on land in a calendar year was painful. You put your kids over this Navy shit and I will always respect you for that. Cajun boy, you my brother. Nothing else needs to be said. Kyle and Sam. To see Black men in leadership positions that genuinely gave a fuck, y'all don't know what that did for me. Thank you for being two of the few bright spots that I could always rely on.

— 7 —

THE REBIRTH

After leaving hell, I returned to a familiar place in Assault Craft Unit 5. I needed home again. I call it home because when I first was there, I grew into a man. As I stated earlier, I was there with OG's in the game. Former drug dealers, former Crips who were still active at the time, all that. ACU 5 was a wild place. This time around, I was the leader, and in a position to change things. I'll come back to that place later, cause we're gonna go into another chapter of mental health. Just know that I didn't wanna leave. When it was time to rotate, me and my detailer had many conversations. For those wondering, a detailer is the person who is in charge of sending you to your next duty station. When it would be time for me to leave here, I would literally have a year left in the Navy. It wouldn't make any damn sense to send me anywhere else. It was a waste of money and time. I'm gonna leave his name out of it, but just know that I learned from him that some people have rocks for brains. Back in the day, we didn't have to go through computers or discuss some shit to an extent. You literally called them up and talked with them like they were your partner from school days. You'd tell them some shit like hey, I see this place is open, or I'm tryna stay here, so send me there. They'll look at it, and be like cool, I got you. A week or two later, your orders were cut. Now, the Navy was brand new, more politicized than ever and flat out just a bunch of bullshit. I'd been in San Diego at this point 17 out of my almost 19 years. I ain't have no plans of going anywhere. San Diego was home. I had literally 12 months left. Just let me finish out my time, get out the Navy hair and I would be gravy. Nah. This fucktard cut me some orders to Guam. Now, do not get me wrong. I love Guam. I lived there for two years of my life, and they were the greatest. I did more partying, drinking and outright wild shit in those two years than at any point in my life. However, as much as I loved it, I wasn't going back. It was a two year obligation, I had less than two to go, and the Navy wasn't about to get anymore than 20 years out of me. See, I'm glad I told y'all this, because this is another mental health stressor that they don't talk about. Time served. Now, I get it. Some people absolutely love this shit. They have fun with this shit. This can be due to a few things. One, they have no responsibilities. No kids, no spouse, no steady girlfriend or boyfriend to irritate the fuck out of them on a daily. Shit. The way they see it is if they can make this money just showing up to a designated place on time everyday, then why not? I couldn't and wasn't gonna do 24, 28, or even 30. More power to them. Unfortunately, not everyone stays that long because they have fun. You got a lot of these cats who stay because they are truly scared. See, they put so much into the Navy. The Navy was the only place in life where they have felt important. The only place where they had a sense of belonging. Truthfully, for many of them, it's the only place where anyone would listen to them. I couldn't be one of those guys. Now, you can call it the bare minimum, but a lot of those guys can't do the bare minimum in their personal lives. That's why Jody visits their house every deployment they are on, and their kids show them no respect. Anywho, back to this detailing shit. He cut me orders and I gave the biggest fuck you ever. I ran my paperwork, which we call a 1306, for failure to obligate the orders. In a nutshell, my Black ass wasn't going. Wasn't anyway in hell. I waited a week or two until the message came back that those orders were nixed. When it did, I was thinking that some sense would ring off in some brains in Millington, Tennessee that hey, keep this dude where he is. Let him finish out his last year, get all his shit together and he can ride off into the sunset. Months went past and I still didn't have orders. I emailed and even called the dude, but he wasn't answering his phone. Either he was that busy, or his panties were in a bunch every time he saw my name. The day after my birthday, that's when the orders came. USS Bunker Hill. Now, why would this dude send me to a ship for my last 12? I knew the answer, so its fuck him right back. I had well over a month until my report date, so I was gonna chill until then. I would literally come into work, show my face and dip. My chain of command didn't care. I was on my way out the door and the way they saw it, I did my job, so they let me have some leisure in my last weeks. If only all commands were like that. My mental health shot up to an outrageous level. I've known Sailors who have been made to work until their last day, and that shit wasn't cool. The military had a knack of pulling shit like that. People had to arrange moves, child care, ship shit off, and because of fucktard leaders, the service member was left suffering without any assistance in the world. The way these so-called leaders saw it, if they had a replacement coming, and they weren't there yet, then the job couldn't be done if we let the service member leave and take care of their personal matters. This was the bullshit that resulted in so many people leaving the service with a huge middle finger to Uncle Sam. Some were stupid enough to put the service before their family. In the end, they would end up with a retirement check, divorce, depression and no purpose. I think the way I went about my 20 years is the reason I look the way I do. Many tell me that I don't even look 38. That's cause I gave a fuck about what I was supposed to give a fuck about. That was myself, my job and my junior sailors. I didn't try to go above and beyond, kiss anyone's ass, none of that. That wasn't me. We already had enough shuck and jive cats in uniform. I wasn't gonna add it to the collection. My bad for going off. Lemme get back to Bunker Hill. I reported on a Friday. September 24th to be exact. My first impression of that ship was a funny one. I had never in my 37 years met a White dude named Marcus. The whole time I was choppin it up with him on email, I thought it was a brother. I got there and it was a White dude. I smile and laugh thinking about it, but he was aight. No issues at all. Now, the crazy shit was that we were going underway that Monday to Seal Beach, California to do an ammo offload. I couldn't complain, cause I had enough rest for the damn near 2-3 months that I wasn't doing shit. Monday came, and that's when I started meeting everyone. It was all normal and good. I still didn't wanna be here, cause it was a waste of time. At least that's what I thought. Then, the Fab Five formed around me and I had no idea that I was about to be Steve Fisher. They were all young, Black, full of life and had their own way of doing shit. Lemme introduce them to y'all. First, there was "The Thug." This was the Chris Webber of them all. Inglewood bred. Naturally on some Blood shit. He held everyone together. He ran things right. He was the superstar. You had my dude Glock. Glock was from Memphis, Tennessee. He was skinny and lanky, but he was like Juwan Howard. When you think Chris was the only one who could bust yo' ass, here he comes. Then, phew, let me introduce y'all to TC 4000. Zone 6 Atlanta official. This was the one. He was the Jalen Rose of the bunch. The gangster. The one with the ultimate swag. The cool, calm and collective one, but could flip and raise hell in a heart beat. I fucked with him heavy. You had Fields, who I always called Chingy, cause he damn near looked just like his ass. He was the Jimmy King of the crew. You know Jimmy could hoop his ass off, but he was overshadowed by everyone else. However, it didn't matter, because the boy still produced. Then, you had KJ. Aww man. We used to roast the dog shit outta each other. He was the Ray Jackson of the bunch. Chi bred. We're gonna come back to him. Upon arrival to the ship, I was told we were headed out to sea for a few weeks. What was cool though was that we were headed down to Puerto Vallarta, Mexico. Not bad, I thought. I hadn't been down there. I knew Mexico was lit because of how Tijuana was back in the day. Now, I was headed to a resort destination. We headed down there the Monday after getting back from Seal Beach. I copped my hotel, which was an all inclusive resort and prepared myself to let my hair down thoroughly. Taking the mini yacht out to land, it was hotter than a camel's nuts in the middle of the Sahara Desert. I needed something to drink. I asked one of the EM (Electricians) homies I had met to let me hit a swig of the gatorade. He smiled as he handed it over and I swigged it. Hot damn. This wasn't gatorade. This was blue ciroc or blue bacardi. It was hard to tell, but it was mixed with some white. This was my welcome back to the fleet moment. We did shit like this in the Navy. Someone always had a bottle on standby. Oh yes, there would be drank. Just make sure you stash that shit in the right place. Trust me, on my 2008 deployment, I was sippin almost nightly. We had our ways. Once we hit land, I took a cab over to my hotel. Once in the room, I just marinated in my own goddess. Here I was, in my last year of military service. What better way to start it off than with five days at a resort destination. I was 37, so hitting every club wasn't on the agenda for me. A good time for me was to find a good book store, some eatery spots, hit the gym and kick it with my people. The first two days were what you would expect. Good eats, pool, a few drinks and just kickin it. I really took this time to myself. The third day, man, this was gonna be one. This was the night of the Wilder and Fury fight. Of course, you know I was cheering for my brother. That's just how it is. Don't like it, oh well. I went over to the resort next to me and kicked it with a few of my engineering co-workers. We were in a big ass pool, with the poolside bar. I didn't plan on drinking too much, but two dollar Jack and Cokes changed my mind. I figured I could have two of them. I took my drinks and just relaxed, and joked in the pool. I hadn't had anything to eat since breakfast, so I didn't want to overindulge because I wasn't trying to get fucked up. About 20 minutes later, I went back for round two. Ok, I told myself, this was it. I can't be getting fucked up and its 100 degrees outside. My homey Oreo, from Chicago, looked at me and I'll never forget what he said.

"You finna be fucked up."

I shrugged it off. I'm a big, solid dude, so I know I would be aight. After about 30 minutes, my head was spinning, but I was still cognizant of where I was at and what I was doing. Ok Joe, I told myself. No more. Relax in this pool and let it wear off. Then, my homegirl hit me with it.

"Get some Long Islands with me."

Fuck, I thought. I couldn't do it anymore. However, here's why I did it. What flashed across my mind is a facetime I got when I was down here. It was my then girlfriend. Lesson one. Never date a broken ass female. That good boy shit of trying to build someone back up ain't worth it. It may work in the movies or rarely in real life, but most of the time, it was just gonna end in heartbreak. My heart was really broken. Her excuse. God told her that she wasn't gonna be my wife. Amazing how women all of a sudden start claiming God told them what to do. Yet when they were blowing me as I was driving through Downtown San Diego, God wasn't anywhere in the picture. I'm not saying I'm perfect, but the imbalance of follow God's order, and we fucking for the last 9-10 months like its nothing is odd as shit. Oh well, mental health challenge of fuck the world was here. I ordered two Long Islands. I don't know what happened over the next few hours, but when I woke up, I was in my hotel room (Unknowingly at first), head slumped on the toilet seat. It was dark and it matched my mood at the time. I got up eventually and staggered my way to the bed where I lay down for a while, trying to collect myself. After a bit, I pulled myself together to head downstairs to one of the 24 hour grill areas. I got myself two hot links, two burgers and ate them like I hadn't eaten in months. Crazy how men are supposed to be so strong and not show vulnerability. Yet here was this woman that I fell in love with. Loved her daughter as if she was my own. Took time with her, did homework and most importantly, taught her the lesson that a man is **ALWAYS** supposed to hold the door open for you, and other shit. I'm not asking for applause, because a fish shouldn't get one for swimming. However, as a man, it's important to show young girls what to expect when they are young so when they are grown, they can tell a real one from someone who is on some bullshit. See, this one here wasn't even about the military. This was about the mental health battle of just being a man. I watched Fury whoop Wilder ass that night, and eventually, I got back to being myself. Lesson learned. Men, never let a woman break you, especially if you did the right things. Now, back to the military side.

When we went back out to sea and it was time to do our engineering shit. Me, Glock and TC 4000 were all on the same headset, clowning our asses off. I mean half the time, we'd end up talking about music, with TC taking no slander about Atlanta rap at all. If they heard the shit we were saying on our channel, they would have never allowed us to be on any operation. As much as I hated being back on a ship, moments like these gave me clarity. It reminded me how I was once in their spot. Now, I was in command central giving the orders, and they were following my lead. It taught me, like mental health, that life is a cycle. Trust, we had real convos as well during these times. It was refreshing to be called UNC. I wasn't being called that shit because I was old. I was being called that because I had garnered their respect. Too many leaders in the military thought that shit came automatically. Here was the mental battle that came with being a good person. The battle of can you lead someone who was once in your place, to the place you are now? The shit wasn't as easy as someone thinks. Especially in the military. I had to learn how to read each individual. Glock was laid back and got shit quick. I'd give him a quick talk and that was all. TC, he wasn't going for no one's bullshit. So, when I rapped with him, I had to break some shit down, and just let him know not to let anyone fuck up his money. I know deep down that his hands matched his attitude and demeanor. You grow and learn to read people. You can spot bullshit from a mile away, and I saw none in him. Inglewood, I knew had the most potential out of everyone. He was the one you could look at and tell in 30 years, he may be the top enlisted person in the Navy. I could talk to him on some street shit. It was the easiest way to get him to understand. People always talk about when you get to the Navy, you let the street shit go. That doesn't apply to those of us who grew up in the hood. Especially, when you are young. Some of your lingo, actions and other shit gone carry over. You'll find yourself calling someone else blood, cuz, folk, depending where you come from. Bruhs speak the lingo of where they are from. This jawn (Jersey), brackin (Cats who came up round bloods), Joe (Chicago), on 4nem (Also Chicago), and for all crazy shit, I don't know who invented the word glizzy. That's how it was. With him, I know he had a low tolerance for bullshit. I may have to talk to him on the side and tell him *"Look my nigga. Don't let these niggas get to you."* Yes, I spoke the N word and still say it to this day in personal convo with my own people. Difference is between now and my younger years, I can think on my feet. I do spoken word, so I know when to enact it and when not to. I can flip it quickly. I know non Black folks may not wanna hear it, but that's how it is. Trust, I served long enough to know every race gotta word they use amongst each other that you bet not hear anyone else say. If this is the part of the book you don't like, I will not apologize. You want real, read on. You want fluff, read the National Enquirer or some shit. KJ, we were from around the same parts. How did we communicate? It started with roasting daily. If there was a day that we weren't joaning on each other, it was because one of us wasn't there. However, I used humor to make things work. In the Navy, HT (welders and plumbers for non Navy personnel) is known as a White boys rate. I'm just keeping it 100. So, as the **ONLY** two Black welders on board, it was my due diligence and responsibility to show him the way. Humor got us started on the right path. It bonded us. Later, it would turn into learning lessons on all sides. I found myself passing down my knowledge to him and everything that I knew. If I didn't know, I wasn't ashamed to tell him. My thing was to make sure he had the tools in a rate that I survived for 20 years. It's already hard enough being Black. It's even harder being Black and people having this impression that you can't do the job. My whole purpose was to have him overcome and handle all the obstacles that I knew would come. Fields, I knew the day would come where I would have to take care of him for all of his old man jokes. There we were, in Seal Beach, CA. Now, don't ask me why Sailors treat Seal Beach like it's a foreign port. It's right outside of Long Beach. It was towards the end of my time on the boat, and he was in the group I was hanging with.

"Ain't it past your bedtime grandpa?"

I took it, giggled and laughed. This was Fields. Always cracking jokes about me being an old man. I took it all night til we hit the Irish bar. I went outside sometime during the night. It had to be almost 10:00 PM. There he was, sitting on the bench, slumped over, throwing up. I couldn't even say anything. All I could do was laugh and get some pics for the gram. The crazy shit is though, between Glock and myself, we took turns being by him to make sure he was good. What's the mental health battle in this you ask? It's remembering how you used to be when you were younger and knowing how to handle shit. Not everything needs to be reported, or as we call it in the Navy "whistle blowing." I was his supervisor and he was good. Nann, nobody from that ship was gonna drag him back and get him in trouble.

I had a lot of trust in people on this ship, but it was always that chance that there was that one. I made sure that there wouldn't be one. By the time I was leaving the ship and they rang me off, each one of them was there, including Glock via facetime from Memphis, where he was with his daughter back home. I didn't know how much I meant to them, til I saw them all there. Inglewood was crying, and that made me shed tears. After a 20 year battle of everything from multiple deployments, divorce, late nights, injuries, surgeries, deaths, I knew I had indeed completed the mission. As I exited the boat, I raised a fist. It was over and complete. For the first time in 20 years, I could be Joe again. Still, Joe's work wasn't complete.

— 8 —

EIGHT MINUTES AND 46 SECONDS

I know, I know. Y'all thought that was the last chapter of the book. How I ended that 20 year journey. Remember, I told y'all at the beginning of the chapter that we had to come back. This is the last mental health battle that I will touch on. Before I go any further, I hope like hell that my experiences in this book have helped someone individually. More so, I hope like hell they will lead to a better military in the future. Especially with some of the shit that I exposed. It's new to some, but very familiar to a lot of us that served. Now, to get everyone riled up in their fucking feelings. I'm pretty sure just by you reading the title of this section, you know where I'm going. Actually, you don't. What you saw on the outside, can't even comprehend what was going on inside the gates of probably every United States military base around the world. 2020 was a helluva year. One, a whole damn Navy ship caught on fire. The BHR (Bonhomme Richard) was cooked extra crispy. That day, I received more phone calls in 24 hours than I had in a whole year. Next, COVID ran rough shots over the world. People were sick, dying, all that. Work had pretty much come to a stand still. I was a 3MC (Maintenance and Material Management Coordinator). In layman terms, I was the head honcho who took all the reports of completed maintenance that was done aboard our hovercrafts. When I went through the school, I thought shit, I'm gonna hate this shit. Turns out I was wrong. It was pretty easy to manage, as long as you stayed on top of your shit. The day was Monday, May 25. Memorial Day. You know, the day when a lot of military people like to get on facebook and post *"In case you thought it was a day to barbecue."* I used to laugh my ass off when I saw that. At that point, 18 years in, there was not one military person I knew that loved, or was absolutely dying to stand duty on Memorial Day, or any holiday for that matter. It was always *"this some bullshit,"* or *"Fuck I'm here on a day off."* Nah. Remember. You love this shit so much, you should be happy to take any duty day serving Uncle Sam. In short, a lot of these folks were, as the kids say nowadays, cappin (lying). I had plenty of holiday duty in my time. It sucked the first few years. This deep in the game, it was nothing. Shit, just the year before, I had duty on Thanksgiving. We stood watch, deep fried turkeys and watched football. The older you get, the more you learn how to make shit work. Anywho, back to the story. I was at home, chillin. I know the beach was packed and I wasn't about to fight through hoards of people. If you saw one pair of booty cheeks in a bathing suit, you've seen 'em all. My only plan was to hit up a BBQ joint later, get some grub and come back home. It was early afternoon, almost evening time when I was eating and caught a glimpse of the news as I flicked through the channels. The police had once again taken the life of another Black man. This one here was different though. This one here had me stopping mid bite. This shit here wasn't a case of police going overboard. Nah. This was flat out **MURDER** caught on camera. It was already an uphill battle being Black in the military. They expected you to just go with the flow. That wasn't me. This really pissed me off and I just knew shit was finna change in a major way. I just didn't know that I would be that actual change. I watched Minneapolis burn that night. I mean, they torched it unlike anything that I had ever seen before. Crazy shit was that my spirit was happy. We were tired of this shit. We were fed up. Being Black should not equate to a death sentence. I thought I was watching a feature film. All I needed was some popcorn. I knew from there, shit was only gonna get worse. Five days later, it did. All throughout the week, they had emphasized at quarters, if you protest, don't go in uniform or affiliate yourself with the Navy. I already knew this. This wasn't the first time a protest happened during my time serving. This shit though, I had to play my part as an elder. My people shut down the 8 freeway in San Diego, near La Mesa, California. I wasn't an advocate for pitting yourself against vehicles, cause the object on four wheels was gonna win every time. However, this time, it wasn't a few dozen. Naw, there were thousands. No car could've gone through that amount of people if they tried. Every Black person I knew in the city, uniformed or not, was involved with this. I found myself talking with many on the phone. Many were expressing going to the San Diego-Coronado Bridge. I was cussin' fuckers out.

"DON'T GO ON THAT FUCKING BRIDGE!!!"

It was one thing to be on that solid ground freeway. It was another to be on a structure over the water, where all they had to do was get a semi to run folks out the way and force them off the bridge. By the end of that night, several structures were burned down in La Mesa, but the wrath had just started. Over the next two months, it seemed like the world had for once gotten fed up as a whole, as I saw protests on television worldwide. That BLM shit? Yea, I knew Black lives did matter, but most of us knew the organization is full of dog shit and cat pis, so we weren't rocking with it, and still don't. Besides, when someone who funds an organization doesn't look like you, that should let you know it ain't for you. Still, the slogan mattered. The CNO (Chief of Naval Operations) would eventually create a group called Task Force One. I guess the purpose of it was to have difficult talks about racism, hatred, all that shit. From my time in the Navy, I didn't trust this shit. I saw so many things like this come and go. For example, When some major shit happened, we usually had standdowns (large group talks) about what happened, here a one team, one fight slogan, and then we were back to the bullshit. The military, to tell you the truth, was all about making shit disappear. If you think professional sports organizations value PR, you ain't seen shit. I'll never forget the day that my command asked me to do a spoken word piece about what was happening. I asked "Are you sure?" They said yea. I had to ask them one more time, because they were gonna let me have free reign.

"Are you sure?" All I remember were the words "McClain, you can do you." With that, I knew that all hell was about to break loose. In a good way for my people. In a bad way, because I was risking my career. I knew what I was about to say. They weren't ready for it and I knew that. It was a Saturday when me and my partner were set to record. We got to the command at around 6:00 AM. I mean, I was setting up this shit to be one of those hit you in the mouth videos. First thing I did was put a bunch of black chairs in a circle. I typed up the names of a bunch of Black people who were killed by an over-excessive police force that were chicken shits with a gun and taped one name to each chair. Now, here is the crazy shit. In the military, we get more training than the damn police. If we have to fire a weapon, do you know how much paperwork and investigative time would come with that? I don't give a damn if it was justified and caught on camera that it was justified. The investigation that would follow would be nuts. In the military, sometimes, when you did the right thing, you knew it could turn into the wrong thing. Not everyone who served holds a weapon. Where I was at, ACU 5, we had certain watches that required it. Before you did anything, you better had known your seven justifications of deadly force and your four universal laws of a weapon. You better remember most of all that you always tried to handle shit at the lowest level possible before you took it to that level. Lemme give you a quick example. If someone was 50 yards away, cussin me out and waving a knife, I didn't need to shoot them. Plain and simple. The deadly force triangle is simple in itself. Opportunity, capability and intent. We knew how to decipher shit. So, why couldn't the police do the same thing? I kept prepping for the video shoot. The camera was gonna switch back and forth from me being in uniform, to me in regular clothes. Those clothes were all black, with the words BLACK WALL STREET on my shirt. I didn't want to make a statement. I wanted to cuss these muthafuckas up top out without saying a cuss word. As we began, I was frustrated. It was hot as camels' balls in the middle of the Dubai desert. The camera team, and I don't know what the technical name of that shit is, they brought the reflector joint to ensure the light was hitting right. In truth, it may have helped the video, but I was cooking. Once I got into the piece, I kept fucking up and getting frustrated. My man, thinking that he was gonna calm me, instead acted like I was in the 4th quarter fucking up with the game on the line. I was pissed, not realizing that he was trying to piss me off more, because he knew the more upset I got, the better this shit was gonna come out. He wasn't a professional with the camera outside of the military for nothing. Finally, on that sixth try, I nailed that shit. Below is the full piece:

One team, one fight, right, yet every night I cut on the news I see that a certain few ain't considered a part of that team, I can no longer take a knee to honor death when those sworn to protect are taking knees to our neck, I can't rest, knowing I got five nieces and if I ever got the call that they were buried in pieces, stand by gone have an entire different meaning, I can't be tough no more, I've cried out for help, my mental is in a 3-1 deficit and most of us are not Lebron James, ain't no comeback from flag draped caskets and 21 gun salutes, but shaving your face seems to be more important than suicide, cause you tell us to suck it up, strap up, we got a mission to complete, hey remember sailor, there's no I in armed forces, yea you right fam, but there is me in that first word, cause my life matters, these Black lives matter, these women lives matter, mental health matters, my family matters but these Carl Winslow types seems to think different when I'm out of this camouflage, and y'all think it only matters when its work related, but you ain't gotta go out those gates and be hated for the bronze uniform that I can never take off, but if I take off, you'll say I'm offset from my military bearing, but try bearing the loss of your brother when he was shot while unarmed, try bearing my mindstate getting pulled over by the laws, try bearing, naw, try lending a shoulder to cry on to your sisters in uniform and then be shooed away, because the last shoulder they cried on said that wasn't rape, or sexual assault doesn't matter, all assault matters, but really only the assault we take to the enemy matters, suck up those tears and cry me a river, but when that river turns to blood...see I am not this uniform, I am tattooed stories of Black men that are no longer here, I am stereotypes and economic inequality wrapped up in a 6'1 package, and I'm not a number or a rank insignia, I am a human being, I was decorated way before I enlisted in the service, multiple purple hearts for the wounds I obtained when they dropped those N bombs, a silver star for the times I put my career on the line to make sure my junior sailors were in their right mind, and my medal of honor comes at a time when my skin is still considered dishonorable in the eyes of society, cause even after 18 years, a clean record, gold service stripes, 5 NAMs, 4 deployments, 3 Battle E's, a 2 year overseas tour and a full ribbon rack, I still remember, that I am forever

Black, and no amount of decoration can terminate the kill on sight declaration they have on my life, so start accepting us for what's under the uniform, cause this cloth can't protect us....

I went off. More so, I was upset because I couldn't cuss like I wanted to. That's the nature of the business. You gotta know your audience. I figured me taking a shot at the current MCPON (Master Chief Petty Officer of The Navy) with the shaves over suicide line, telling how fucked up the Army was about how they covered up Vanessa Guillen's murder (the assault lines), would get them going. We spent 15 hours that day between live filming and the studio, getting everything right. Now, it was a wait until everything was put together and sent off. I was ready for my phone to be tapped after this shit. A couple of weeks went by and the video was finally done. This next part is exactly what I was told. Who told me ain't the concern. Just know, I believed every word. The head folks at my command enjoyed it. It got sent to the people above us. It wasn't to their liking (I'm being nice saying this). They sent it off to the people above them. Same shit. When it got back that they wasn't fucking with it, some of my folks sent that shit to DC. Who sent it? I don't know (I actually do). Who received it? I don't know. What I do know is that a little while later, I was called to head to someone's office. A Captain left me his number and told me to call him. I ain't know who homey was. I called and he ended up being one of the officers on the task force. A brother. I wasn't a rank on this conversation. I was just Joe. He loved every ounce of that shit. Okay, cool. Then, not too long after that, maybe, a few days later, I got a message on the book. It was one of my people's on the East Coast. They asked for my number. No issue. Whole time I'm thinking that fam was about to holla at me, I got hit with "Force Master Chief" wanna holla at you. Force Master Chief? The brother? Aight, either two ways this was gonna go. A brother was finna get cussed the fucked out. Or, a brother was finna get chewed out, without cuss words. Force Master Chief didn't just ask for people's numbers through other folks. I was on the quarterdeck when he called. For the next hour, I was in the parking lot, sitting in my car. Laughing, shooting the shit and talking about real shit with a highly respected member of the armed forces. Man, I don't know who the video got sent to, but I had some of the most powerful Black men on my side. If it couldn't get any better, I was then told that one of the top Admirals in the Navy was gonna be in Daygo the following week, and he wanted to sit down with me one on one. I had swung for the fence with this one and hit a Joe Carter, 1993 World Series clinching home run. The Admiral ended up not coming, but I got more news. I was told to get my house in order, because I was heading to DC in September. Shit was real now. Was I excited? Indeed. The Black voice that many didn't wanna hear was coming. Also, I had something in DC that I was trying to kick it with, so that was cool as well. Where was the mental battle you ask? The battle to be Black. Point blank period. It wasn't about me at this point. It was about me making this shit easier for all the young brothers that were gonna come behind me. When I came in, I had OGS who took me under their wing, but even we weren't dealing with shit like this back in 2002.

We knew racism was prevalent, but social media and the exposure it now had took it to an astronomical level. One day, when I first dropped that Black History piece at my command that went viral in 2020, one of my partners hit me up.

"Go to this military page man. Just look at it and hit me back."

He gave me the name of the joint and I searched it out. Of course, it was one of those typical military pages where we crack jokes, share memes, all that goofy shit we do. However, after a quick scroll, I saw my video. Then, I saw the comments. Now, this was every rank, race and creed on here. It was active duty folks and veterans. I lost count of how many N bombs I was called. Porch monkey, fucker and my favorite, "Why do they still complain? They can serve." I remember that one vividly. Who knew that one piece would lead to the shit that I was involved with at this current moment. The crazy shit about that piece was that I flew out the next day to go perform in Tulsa, on Black Wall Street, at a play that was being held. I didn't even know it was being recorded. I woke up to text messages saying go to facebook, you went viral and nigga you finna piss these White folks off. That shit was wild. In the end, it helped me at the right time. Not for glory or any of that, cause I've been in the world slam, this slam, that slam, won poetry slams, performed in front of thousands, that ain't shit. It gave me balance on where I was at that moment. Tulsa was a lost story in history. Let me correct myself. It wasn't lost. White folks just tried to hide what the fuck they did. So imagine thinking damn, I am a part of the same military that once bombed my people? This being Black shit can be conflicting at times. Never conflicting against who I am. Never that. But, just some of the shit we represent. I mean, do you think all the bruhs wanna have the USA across their chest during the Olympics, knowing that they are only cared for because they can shoot a basketball? That's another mental battle in itself. It's difficult at times, especially when you've grown into yourself and realized who the fuck you really are. No one is grown by 18. Not even 21. If you ask me, I didn't grow up until I was 28, and I still had additional growth to accomplish. Now, at almost 20 years in and closer to 40, I was there, with still much to learn. However, willing to sacrifice your career so late in the game, that was the biggest growth of all. No one could tell me shit. I ended up not going to DC, because 4-5 (We don't say his name round here) shot down all diversity training. However, I still left my mark. Over the next few months, I would lead or assist in talks with hundreds of sailors. Some shit, didn't surprise me. Other shit, it made me look at folks a whole different way. Still, I accomplished what few had done. I pissed the MCPON off. I pissed brass in San Diego off. Yet and still, here I was. Still thriving and existing. On my daddy, you'll never hear any sailor that was ever under me say that I fucked 'em over. My mind may be fucked up from dealing with the issues my skin color brings on a daily. However, ask me how my mentality was when it came to being there for my fellow Black sailors, and it was on level 10,000. Born Black, die Black, embrace Black. That'll never change and I'll never fold....

ABOUT THE AUTHOR

Joe Rainer McClain Jr. is a 20 year, retired Navy veteran. During his career, he served aboard multiple platforms. USS Nimitz (CVN 68), ACU 5, USS Peleliu, FLC San Diego, USS Frank Cable, Naval Medical Center San Diego, USS Harpers Ferry and USS Bunker Hill. He received numerous citations and awards, including six Navy and Marine Corps achievement medals. He participated in Operations Iraqi Freedom, Enduring Freedom, Piracy Ops and spent time in the South China Sea later in his career. Overall, his Naval career saw him embark on four deployments, visiting 27 countries on five continents. Outside of the military, he is a dynamic speaker and author. He placed Top 25 in the world at the 2019 Individual World Poetry Slam, is a multiple time poetry slam champ and has toured nationally, headlining numerous events in cities across America. As an author, this is his 11th book. His works have been acquired by some of the biggest names in sports and entertainment. His passion, however, is speaking on stage about overcoming the obstacles in your life to be the best you that you can be.

Famous
BLACK
HISTORY

Made in the USA
Monee, IL
04 November 2026

67383423R10069